ADVANCED COMPUTER AIDED MANUFACTURING APPLICATIONS

CNC TURNING

Pavel Ikonomov Ph. D.

Preface

This book is created to help users of various 3D CAM software and CNC machines to create programs for CNC machines. Major topics are programming of CNC machines using standard G and M code command. Each command is explained in detail and presented with detailed subsequent images for each small step that helps reduce possible misinterpretations. An effort was made to explain command, programming sequence, and requirements while keeping the description to the minimum.

In my experience teaching, CAM CNC machining was the most difficult part in any Computer Aided Manufacturing, Computer Integrated Manufacturing process and CNC programming laboratory we created for any single command when more than one CAM software system or CNC machine were used. This tutorial was originally written for students in the College of Engineering and Applied Sciences to help them with a group of courses: Introduction to Machine processes - EDMM 2540, Computer Aided Manufacturing - EDMM 3580, Manufacturing System Integration – EDMM 4580, Concurrent Engineering - EDMM 5460, and Computer Aided Manufacturing Applications - EDMM 6580.

EDMM 4580, EDMM 6580, and EDMM 5460 are advanced courses taken by senior undergraduate and graduate students and require extensive CAD /CAM manufacturing competence. EDMM 2540 and EDMM 3580 are attended by students with a broad range of CAD/CAM experience; some with extensive, others with only the introductory programming experience and computer skills. This tutorial is easy to follow even for a user with limited CNC programming experience while at the same time providing support to advanced users.

Detailed descriptions of all the possible program situations with different CNC machines within a single volume of a book is impossible. For any additional information, users can refer to operation and programming CNC manuals, CAM software manuals, help files, and on-line help from CAM vendors. Also, notice that different versions and license schema by CNC or CAM companies may show different result from the one explained in this tutorial.
The chapters in this book are not meant to be followed sequentially, so please use the chapter you need to a specific command and programming method. This book is **not a reference** guide and is intended to give sufficient (but not complete) descriptions of manual and CAM-based CNC programming.

Almost all of the examples shown in this book were individually created. Any similarity of objects, parts, and drawings used in this tutorial are coincidental and unintentional. All files created for this tutorial will be available as a free download on a Web site specified by the author.
Access to the site with all the free files from the tutorial need be requested from the author. Due to the file sizes, E-mailing of the files is not practical.

Books samples for each chapter are available at web site:
http://homepages.wmich.edu/~pikonomo/

About the author

Dr. Pavel Ikonomov is Assoc. Professor in Engineerin Design, Manufacturing and Management Engineering Department at Western Michigan University.

He earned his bachelor degree from the Technical University of Varna and his first master's degree at M.E. in Mechanical Engineering and Manufacturing Technology from Technical University of Varna. His second master's degree was earned from Muroran Institute of Technology, Japan, and his Ph. D in Precision Manufacturing Engineering from Hokkaido University, Japan.

Dr. Pavel Ikonomov has worked several years as chief mechanical engineer in a petroleum company and Asst. Professor at Technical University of Varna - Bulgaria. Later he held positions as CTO at Virtual Reality Center Yokohama - Japan, Associate Professor at Tokyo Metropolitan Institute of Technology - Japan, Visiting Professor at UCLA and National Institute of Standard and Technology (NIST). He has extensive industrial and teaching experience in different countries, university research centers, and companies. Dr. Ikonomov has contributed significantly to the development of new data exchange standard working on establishing of the STEP standard at Hokkaido University, Japan and information exchange between design applications and the virtual environment at NIST. He is considered an expert in CAD/CAM, Virtual Reality simulations for industry and nanomanufacturing, and 3D printing. Dr. Ikonomov had published more than 140 papers in journals, proceedings, chapters in books, and have several patents (last two in 3D metal printing.)

Email:
pavel.ikonomov@wmich.edu
pavel.ikonomov@gmail

Table of Contents

Preface	ii
About the author	iii
Acknowledgments	vii
Table of Contents	v

Chapter 1
Introduction to Computer Numerical Control machining — 1

Why NC?	2
What is NC?	
How NC machining works?	
History	2
CNC Programming Language development	4
NC and CNC	5
Direct Numerical Control	6
Distributed Numerical Control	7
CNC Standards	8
Conversational programming of CNC	8
Programming CNC using CAD/CAM	9
CNC machine and controller manufacturer	9
References	10

Chapter 2
Elements of CNC machining — 13

CNC system elements	14
CNC control system	14
CNC interpolation	16
CNC Machine Control Unit	17
CNC Software	18
CNC machining process flow	19
	14

Chapter 3 — 23
Fundamental concepts of CNC machining

Axis Motion	24
Coordinate system	24
Turning CNC machine fundamentals	27
Programming in diametrical/radius values	29
Programming with absolute coordinates	29

Programming with incremental coordinates	31
Tools for Turning	34

Chapter 4
CNC Turning Programming

41

CNC Turning programming	42
G and M command used in CNC turning	44
G00 Rapid Linear Motion	46
G01 Linear Interpolation	47
G02 Circular Interpolation Clockwise (CW)	49
G03 Circular Interpolation Counter Clockwise (CCW)	51
G04 Dwell	53
G20 Inch Units	55
G21 Millimeter Units	56
G28 Return to Home Position	58
G29 Return from Home Position	58
G32 Single Thread Cycle	60
G41 Tool Nose Radius Compensation Left	62
G42 Tool Nose Radius Compensation Right	64
G40 Tool Nose Radius Compensation Cancel	66
G54-G59 Select Work Coordinate System	68
G70 Finishing Contour Cycle	70
G71 Rough Turning Contour Cycle	71
G72 Rough Facing Contour Cycle	73
G74 Peck Drilling Cycle	76
G75 Grove Cycle	77
G76 Threading Cycle	79
G90 Absolute Programming	81
G91 Incremental Programming	83
G96 Constant Surface Speed	84
G97 Constant Spindle Speed	86
G98 Feed Rate Per Time	88
G99 Feed Rate Per Revolution	90
M00 Program Stop	91
M01 Optional Program Stop	93
M02 Program End	95
M03 Spindle Clockwise Rotation (CW)	96
M04 Spindle Counter Clockwise Rotation (CCW)	98

M05 Spindle Stop	99
M08 Coolant Start	100
M09 Coolant Off	101
M30 Program End and Reset to the Beginning	102
/Block skip/	103
(_ _ _ _) Comments	105
Chapter 5 **CNC Turning Examples**	109
Part 9003	111
Part 0345	115
Part 1337	121
Part 9002	125
Part 0010	128
Part 9001	132
Part 0001	138
Part 2001	146
Part 9004	154
Part 1101	161
Part 9006	164
Part 9007	167
Part 0009	170
Part 9008	176
Part 0019	180
Part 1001	183
Part 1102	191
INDEX	201

Chapter 1

Introduction to
Computer Numerical Control machining

Why NC?
Why NC?
What is NC?
How NC machining works?

Since it appears NC machining has changed the way we machine products. Initially designed for quality and quantity they have become an indispensable part of any production process, from a fully automated system as Flexible Manufacturing System (FMS) for multiple variation of parts and products to a single workshop production of unique or customized products. With the advance of computers NC technology itself has developed and become more sophisticated in hardware and software usage while at the same time user interface, the why NC operator deal with NC machine, have been simplified. As a result, complicated parts can be produced using general CAM software. Today almost every significant CAD software company offers CAM integration package as well. The number of NC machine or CNC as they are called nowadays has increased steadily. Typically, NC produces up to 60% of overall machining operations even though they are about 20% of the total number machine of machine plans equipped. The standards developed and easy data exchange has contributed considerably to make the CNC machine one of the most widely used means of machining in the global market. The internet has made the CNC production fast easy and convenient. You may have a design studio in Detroit, send the CAD design to Mexico, produce the NC program and sent it to Asia, machine the part there and ship to the USA or other markets.

History

Applications in the early stages of Industrial revolution necessitate the introduction of the form of NC machining. Early effort to automate production was a simple use of pulleys, belts, cams, etc. For example, during Middle Ages, some churches used rotating drums with preposition fixed pins to control chimes. A type of NC utilizing punching card for creating a range of shapes and patterns in knitting machines was used in England early in 18th century. Later in 19th century automatic playing piano "player piano" was invented with keys move following a pattern of holes in a punched paper scroll. Although the advantages of the automation the cheap manual labor and better quality was a reason to use human to control machines.

During the World War II, a shortage of qualified workers and the increased requirements for quantity and quality industry recognize that there is need of new technology. The war machines such as airplanes, tanks, guns, required high-quality identical parts and in high volumes. Highly qualified machinists could produce high-quality parts, but could not meet the manifold increase of quantity. Due to the necessities to war battles, US Air Force needs many new high-quality identical airplanes to be manufactured with high quality. Several companies have been chosen to develop and manufacture numerical control systems to meet this demand.

The special requirement that NC machines need to meet were:
1. Increase production output together while guarantee high quality and accuracy of parts being produced

2. Stabilize manufacturing cost while parts are manufactured quickly
3. Producing complex parts that were not possible by the conventional manual methods

Figure1.1 NC design was intended to meet the production demand during the World War II

The parts produce with NC machine were not aimed to a mass production rather flexibility to meet certain production requirement. The production quantities of parts are for small and medium-size batches, with varied sizes and geometry that can be produced with similar steps and settings, see Figure 1.1.

During World War II the world's first digital computer Electronic Numerical Integrator and Computer - ENIAC was designed and built to calculate artillery firing tables for the U.S. Army's Ballistics Research Laboratory. Project PX was constructed by the University of Pennsylvania's Moore School of Electrical Engineering from July 1943. It was unveiled on February 15, 1946. It contained 17,468 vacuum tubes, weighed 30 short tons (27 t), and consumed 150 kW of power. In early 50's MIT developed more advanced vacuum tube computer "Whirlwind" that operated in real time, used video displays for output, and was thousand times faster in computational instruction than ENIAC.

In 1948, "Parsons Corporation" of Traverse City, Michigan, was awarded a contract from US Air Forces to make tapered wings for military aircraft. Earlier, John T. Parsons (Detroit, October 13, 1913) pioneered numerical control for machine tools in the 1940s. Together, with his employee Frank L. Stulen, they were the first to use computer methods to solve machining problems. It was used for the accurate interpolation of the curves describing helicopter blades.

MIT's servo Mechanism Laboratory was subcontracted in 1951 and later took over the total NC development. The first working NC three-axis numerical controlled machine tool was created at MIT in 1952 and shown to the military, the aerospace industry, the machine tool industry and the technical media, see Figure 1.2. MIT's Whirlwind computer was used to control simultaneously three-axis movement on a retrofitted Cincinnati Micron Hydrotel Vertical Spindle milling machine. In few years most of the machine tool manufactures start production of NC machines.

In 1955 a standardization of Numerical Control system was recommend to Air Force by the subcommittee of Airspace Industries Association (AIA).

Figure 1.2 The first 3-axis numerical controlled machine – MIT

CNC Programming Language development

G-code developed by the Electronic Industries Alliance (EIA) in the early 1960s is a common name for the programming language that controls NC and CNC machine tools. The final revision of the standard RS274D was approved in 1980. Most NC/CNC machine tools today can be run using programs written in RS274. However many machine tool companies introduced different implementations, modification, or addition of the language. As a result, a CNC program that runs on one machine may not run on another, without modification, from a different tool maker.

The newly developed RS274/NGC[1] language has significantly enhanced the program capability further than those of RS274-D.

To simplify the NC programming for complex parts MIT start developing computerized Automatically Programmed Tool (APT) 1954. It was based on the concept to use English-like words (so-called high-level) language that generates instructions for CNC machine from geometry description and tool path motion. Subsequently, new versions of APT, APT II in 1968 and APT III in 1961 were released. APT IV version supported complex surface definition for machining. As it evolved APT became a standard for CAD/CAM CNC system. Further developments of APT for smaller computers were introduced: ADAPT (IBM), MINIAPT (West Germany), FAPT I and FAPT (France). Today there are virtually hundreds of programming languages, tools, and software based on APT or developed independently.

Table: APT programming example[2]

```
$$ DEFINE SCULPTURED SURFACE TO BE MACHINED
P1=POINT/0,0,20
P2=POINT/30,-5,26.5
P3=POINT/60,-5,26
P4=POINT/86,0,20
P5=POINT/-6,-30,15
P6=POINT/28,-25,22.75
P7=POINT/66,-25,22.5
P8=POINT/100,-30,15
SPLINE,P5,P6,P7,P8,      $
SPLINE,P9,P10,P11,P12
R1=POINT/30,-15,50
R2=POINT/70,-15,50
R3=POINT/0,0,0
R4=POINT/0,-30,0
C1=SCURV/CURSEG,R1,R2
C2=SCURV/CURSEG,R3,R4
DS1=SSURF/TRANSL,C1,CROSS,C2
TA=VECTOR/0,0,1
CUTTER/10,5
FROM/(STPT=POINT/50,-60,50)
SCON/INIT,ALL
SCON/DS,DS1,PARAM,0,1,0,1,ON,NORMAL
SCON/PS,TO,PS0,MINUS,0
SCON/AXIS,TA
SCON/STEPOV,0.5,5,0,0
SCON/FEED,100,200,50,3000
SMIL/ZIGZAG,DS,PARAM,0,0,TANSPL,PLUS,STEPOV,PLUS,0
```

NC and CNC

At present, Numerical Control NC (Numerical Control) and CNC (Computer Numerical Control) term carried the same meaning. By NC it is assumed CNC Numerical Control which controls the automation of machine tools with sophisticated computer control and memory capabilities and inherent all control capabilities of the old NC machines. The CNC machine tool uses servo motors and feedback technology to perform the multi-axis motions. The CNC program, lines of (ASCII) text, specifies the motion, control functions and coordinates of points is prepared from CAD file (automatically using CAM software or manually) and inputted to the CNC memory manually, with a floppy disk or downloaded using serial/network connection. CNC has enough memory to store multiple programs. Further, the program can be recalled from memory executed, modified and uploaded to storage on computer or mainframe. The program remains in memory and can be reused and modified until is explicitly deleted.

Table: Differences between NC and CNC machine tools

NC	CNC
Hard wired –Can' be changed	Computer controlled software based –easy changeable
Fixed Control logic for code, functions, commands	Microprocessor resident control logic - changeable
Control logic can be changed only with circuit boat replacement	All programs changeable
No memory and storage	Memory storage
Programs can't be created or modified on the machine. All programs are created externally on punched tape	Programs can be created or modified and stored on the machine. Program inputted via the network, serial communication, floppy disk (FD) or manually created.
Programs stored only on punched tape	Programs can be stored on mainframe, server or CNC memory and uploaded trough directly connection or media (FD)
Each part need separate tape	Memory capacity store multiple programs indefinitely

Direct Numerical Control

The concept of Direct Numerical Control (DNC) is linking a computer to the CNC machine CPU directly, see Figure 1.3. In 1960-s one of the first companies to implement DNC were Cincinnati Milacron and General Electric. In 1970's many DNC machines were built based on the concept of usage of one networked (shared) main computer to program, control and execute machine operation on several NC machines. Although promising the DNC systems were used only to transmit the program to CNC (NC) machines. Today is not very often used, although virtually all CNC machines are capable of DNC trough serial, network communication, storage cards, and devices.

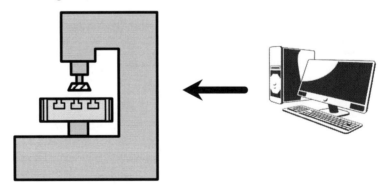

Figure 1.3 Direct Numerical Control

Distributed Numerical Control

With the development of NC machines and computer technology, new types of Computer Numerical Controlled (CNC) machines were built. They have a computer was placed directly on the machine. At present virtually all NC machines are CNC and the new NC is used inter-exchangeable with CNC.

The capability of CNC machine to control and store the program made possible stand-alone operation. With the rapid advance of computer technologies and networking and CNC machine a new type of control, so-called Distributed Numerical Control was developed, see Figure 1.4. CNC and computer were connected through the network that provides not only programs communication (downloading/uploading) but also possibilities to control and automate the whole manufacturing system. Thus DNC processed can provide functions such as machine and robot monitoring and control, scheduling and balancing of the automated lines, control of the status data and managing of the whole process.

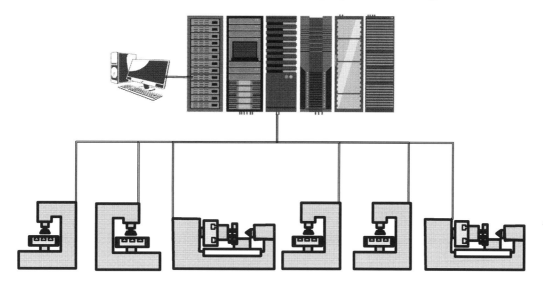

Figure 1.4 Distributed NC

CNC standards

At present, there are two widely recognized and accepted standards for language for computerized numerical control (CNC) programs. International Standard Organization (ISO) developed ISO 6383 and the Electronic Industries Alliance (EIA) developed EIA RS274D. Both standards are very similar and are followed by most of the countries of the word.

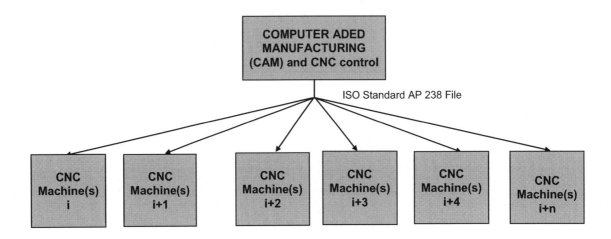

Figure 1.5 AP-238-STEP- Boeing

At the time of developing of those standards (40-50 years ago), paper tape was most common data medium and CNC machines could process only simple commands. With the development of computers processing and graphics capabilities 3 dimensional (3D) data is easily handle and process, still NC programming stayed behind as the only one part of in CAD/CAM process that is not capable of 3D processing. To solve this problem the new Application protocol AP-238 as a part of ISO STEP (STandard for the Exchange of Product model data) standard was developed, see Figure 1.5. It allows the information for control a CNC machine and 3D data from CAD/CAM system to be associated and create CNC control file (completely documented). An AP-238 file is a capable caring product, manufacturing/machining, as well as processing information (including tolerance requirements), thus virtually encapsulating all the information needed for production.

Conversational programming of CNC

CNC controllers can be fitted with conversational (nonstandard) programming capabilities. The implementation is proprietary and depends on each CNC machine maker. Each machine may support both standards EIA/ISO (G and M codes) and conversational programming in addition. The intent is to be easy for use on the machine floor by operators for a single part, manufacturing feature or small batch production. Many of the CNC controller software can also convert the conversation program to standard standards EIA/ISO codes. Some of the most advanced conversation programming systems can be effortlessly programmed not only for simple parts but for parts with complicated features.

8

Programming CNC using CAD/CAM

Modern design and manufacturing processes have been applying the most advanced computer technology. The idea about applying digital model workflow is to use same CAD/CAM 3D data model for both design and manufacturing process, thus enabling effortless updating for the data model for any changes during the production flow. 3D geometrical model from CAD is used in CAM to produce cutter path, including cutter offsets, tool selection, cutting speed and feed rate. Before transferring to the CNC machine the cutter path program is verified for a specific machine for tool collision, shape, precision, etc. CAM software allows programming of complex parts and 3 dimensional (3D) shapes that are not possible with manual programming. Finally, at the inspection phase of the manufacturing process, the 3-D geometrical model from CAD can be used to compare the data from the measured part from the machined part.

CNC machine and controller manufacturers

Main controller manufacturers are Fanuc, Bendix, Bridgeport, Cincinnati Milacron, General Electric, Giddings, Haas, Lewis, Mitsubishi, Okuma, Siemens, and Yasnak are using sometimes their own CNC controllers with a specialized machine processing unit (MPU) or enhanced personal computer processor units. Virtually almost all CNC manufacturers follow the ISO/EIA programming standards for the most of the function that makes CNC programs portable between different machines, see Figure 1.6. On the other side, existing CAM software post processors (including their own post processor CNC code generators) support creating programs for most of these controllers.

Figure 1.6 Different CNC machines

The major CNC tools manufacturing companies are Fanuc, Bendix, Bridgeport, Cincinnati Milacron, Emco, General Electric, Giddings, Haas, Lewis, Milltronics, Mitsubishi, Okuma, Siemens, and Yasnak. At present there are thousands of CNC companies making CNC machines for milling, turning, grinding, water, laser and plasma cutting, punching and nibbling, electrical discharging machine – die sinking and wire types, etc.

References:

[1] National Center for Manufacturing Sciences; The Next Generation Controller Part Programming Functional Specification (RS-274/NGC); Draft; NCMS; August 1994.

[2] Example of a regional milling program, retrieved on 04/04/2017
http://www.catapt.com/APTssman/apt2_8.htm#C8_4_2

Chapter 2

Elements of CNC machining

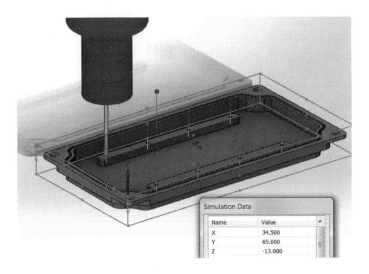

Figure 2.1 Point to point drilling operation

CNC system elements

CNC control system

There are two modes of CNC control system: Point to point and Continuous path. Point to point (PTP) is de-facto a positioning system. Tool move from a point to next programmed point, then performs an operation such as drilling, boring, tapping, reaming, and threading, see Figures 2.1, 2.2, and 2.3.

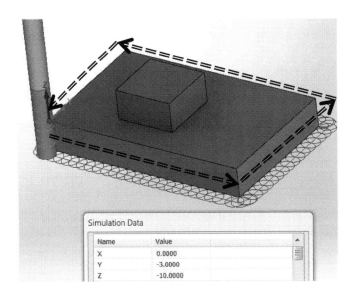

Figure 2.2 Point to point milling operation

The direction of travel can be along one of the X and Y axis axes (Figure 2.1 and 2.2), or 45° degree angle path (Figure 2.3).

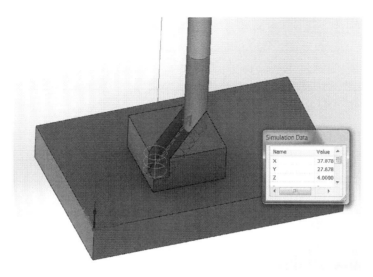

Figure 2.3 Point to point milling 45°

Typically the PTP control system operates in the predefined steps. For example, as shown in Figure 2.1, the initial step is positioning to predefined holes center coordinates; the next step is the actual machining along Z axis with specific rotation speed, controlled feed rate, and depth of the cut. Subsequent steps are rapid retracting and repositioning to the next point, followed by repeating the machining, the final step is rapid retracting to a predefined final position.

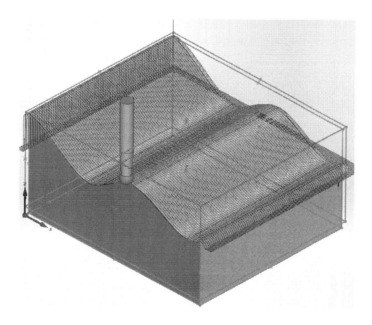

Figure 2.4. Continuous path contour 3D machining

The continuous path, also called contouring system, is synchronized motion on the predefined path. The most common paths are linear and circular arc motion. The continuous path system involves simultaneously control on two, three or more axes, see Figure 2.4. Depending on the capabilities to control simultaneous the continuous path on several axes, CNC machines can be classified on 2, 3, 4, 5, and 6 axis types. The simultaneous control of those axes is very complex and involves capabilities of those machines to control the motion of driving motors independently at various speeds. The contouring on a specific path is done by interpolation.

CNC interpolation

Interpolation is precise movements of the tool CNC on different axis while keeping the tool precisely to the desired programmed path.

Linear interpolation
Linear interpolation is a movement of CNC tool on linear path calculated by the CNC controller. For example, X2.0 commands the machine to move 2.0 only on the X axis, Y1 commands the machine to move 3.0 only on one Y axis. X4.0 Y4.0 the machine to move on X and Y with tiny single axis increments simultaneously on each of the axes thus creating a 45° line, see Figure 2.5. Similarly, a CNC machine can move precisely to any point defined in the program (e.g. next point X5.0 Y5.0). When linear interpolation command is issued only end point and feed rate (speed command) must be given, the CNC machines know its existing position, so it can calculate the path to the new point. Movement on each axis is controlled by CNC electronic device for up to five/six axis simultaneous (linear axes X, Y, and Z; rotation axes A, B, and C). Most common CNC machine can move in two or three axes simultaneously. Linear interpolation can be used to approximate different types of tool paths required in CNC machining: circular/arcs, spline curves, helical, parabolic, etc. Most of these approximations are calculated using CAM system in advance and resulting path programs is transferred to the CNC controller. This makes possible machining of complex 3-D surfaces such as car body shape, ship hulls, aircraft wings, etc.

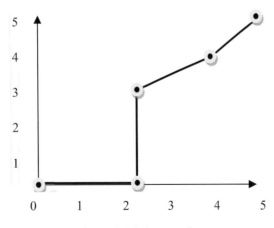

Figure 2.5 Point to point

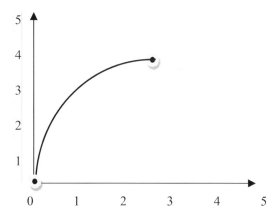

Figure 2.6 Circular Interpolation -continuous path

Circular interpolation

Circular interpolation is a movement of CNC tool on circular path calculated by the CNC controller. For example, X3.0 Y.3 R1.5 commands the machine to point move on an arc with radius R1.5 trajectory (assuming starting point X1.0, Y1.0), see Figure 2.6. Movement on each axis is controlled by CNC electronic device for up to six axis simultaneous (linear axis X, Y, and Z; rotation axis A, B, C). Typically, CNC machines support circular interpolation only on two axes simultaneously, for a circle/arc path lying on the same plane (XY, XZ, or YZ), but some more advanced CNC controllers can provide circular interpolation control in any direction.

CNC Machine Control Unit

Machine Control Unit (MCU) controls all CNC functions such as data processing, input, output, and I/O (Input/Output) interface.

Typical MCU includes the following devices and systems:
1. CPU (Computer Control Unit)
2. Memory- RAM/ROM
3. Storage (secondary)
4. Communications
5. Spindle speed control
6. Servo drive control
7. System Bus –interconnect all the systems
8. Programmable machine controller (PMC) or sequencer

CPU

The Central Process Unit (CPU) controls the components of MCU using programming software loaded in the memory. CPU has three main sections: Control section that coordinates and controls all the functions executed by the CPU and access instructions from the memory; arithmetic-logical unit (ALU) that carry out all calculations and logical operations; and immediate access memory that store internally data from ALU and all instructions for immediate execution.

Memory
Includes Read Only Memory (ROM) and Random Access Memory (RAM). ROM contains the operating system and its permanent memory, remains in the storage even after the powers is switch off. NC programs are stored in RAM only for execution and are removed when the power is off.

Storage
Floppy disk or hard drive, similarly to use on a personal desktop computer, can contain all programs need for NC machining. In the past, punched paper tapes were also used as a storage device. Modern CNC machines can store all the information including control operation. In addition, most of the new CNC machines can access program libraries as well as store programs on the remote server through local area network. Flash memory or non-volatile (no power need to store information) computer memory (also called bubble memory) can also be used for storage functions.

Communications
Communications include: interface I/O (input/output), cathode ray tube (CRT) or liquid crystal display (LCD) interface, RS232 serial communications, storage interface, network interface. Communication between CPU and CNC components is transported mainly through the system bus.

Spindle speed control
Most CNC machine can control the spindle speed is using a special speed function in the part program. The spindle speed control consists of serial speed control circuit and spindle speed feedback interface.

Servo drive control
Convert the machine control pulses to drive the axis control motors. Control from CNC interface is done with the low power electrical signal that necessitates servo drive amplifiers that directly drive servo motors. Feedback system detects the actual position and sends the information back to CPU allowing precise regulation of the position and velocity.

Programmable machine controller (PMC)
Programmable machine controller provides to the following functions: Automatic tool change, Coolant control, limit (end) switch interface, timer and counters, CNC input/output interface and others.

CNC Software

The computer in CNC operates using three types of software:
   1 Operating system software,
   2. Machine interface software, and

3. Application software.

The operating system software is the main program for execution of CNC functions. There are few existing operation systems software vendors that provide it together with the control unit. The operating system software interprets the CNC part programs and generate the corresponding control signals to drive the machine tool axes

The machine interface software, provided by the machine builder, set the communication between the CPU and the machine tool to accomplish the CNC auxiliary functions.

The application software consists of the CNC part programs for machining applications and other functions at the end user place.

CNC machining process flow

Creating of the CNC program is the final step of the CNC manufacturing process.
CNC manufacturing process = conventional manufacturing process.
Prior to start machining a complete manual or computer-aided CNC manufacturing process need to be developed.

Manual CNC process flow:
1. Develop or obtain the part drawing.
2. Decide which machine(s) will perform the operations needed to produce the part.
3. Decide on the machining sequence and decide on cutter-path directions.
4. Choose the tooling required and their organization on the machine.
5. Do the required math calculations for the program coordinates including tolerance requirements.
6. Calculate the spindle speeds, feed rates, and depth of the cut required for the tooling and part material.
7. Write the CNC program.
8. Prepare setup sheets and tool lists (these will also be used for manufacturing operators).
9. Verify and edit the program, using a virtual machine simulator such as CNCez, NC plot, Inventor HSM, etc.
10. Transfer the program to the CNC machine.
11. Verify and edit the program on the actual machine and make changes to it if necessary.
12. Run the program and produce the final part.

Computer Aided CNC process flow:

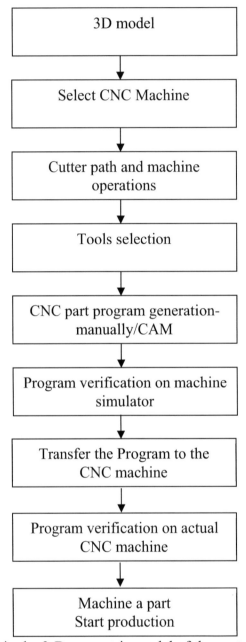

1. Develop or obtain the 3-D geometric model of the part-CAD
2. Decide which machining operations and cutter-path directions are required to produce the part (computer-assisted or from engineering drawings and specifications).
3. Choose the tooling to be used (sometimes computer assisted).
4. Run a CAM software program to generate the CNC part program, including the setup sheets and the list of the tools used.
5. Verify and edit the program, using a virtual machine simulators such as Mastercam/SolidCAM/InventorHSM/SurfCAM/Cimatron/CATIA/Siemens NX, etc.

6. Download the part program(s) to the appropriate machine(s) over the network and machine the prototype. (Sometimes multiple machines will be used to fabricate a part.)
7. Verify the program(s) on the actual machine(s) and edit them if necessary.
8. Run the program and produce the part. If in a production environment, the production process can begin.

Chapter 3

Fundamental concepts of CNC machining

Axis Motion

All existing CNC machines are based on the standard so developed programs can be run on any machine, since the motion control and coordinate points are defined same way. A CNC program is based on the coordinate system that is common for any machine independently of the axis motion implementation by different manufacturers. In some cases, the tool move, in other the table moves or any combination of both. Yet according to the programming concepts, based on standards, it is always assumed that the tool moves relative to the workpiece and the tables. Thus, the programmer can create a tool path movements for tools regardless of the way machine works.

Coordinate system

The most common coordinate system used in CNC machining is the rectangular Cartesian coordinate system. CNC tool position is controlled related motion to this coordinate system along axes. Each linear axis is perpendicular (90° angle) to others. One can find the positive direction of an axis of the Cartesian coordinate system using right-hand rule. The Cartesian coordinate system is also called right-hand coordinate system and both terms are used regularly.

The direction of each axis can be determined by holding right hand with extended thumb, forefinger and middle finger perpendicular to each other. As shown in Figure 3.1, the thumb finger shows the positive X direction, forefinger the positive Y direction, and middle finger the positive Z direction. The machine spindle with the tool is always in the Z direction.

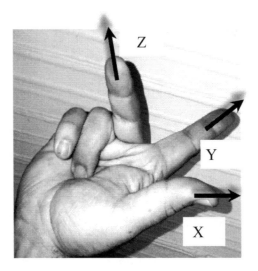

Figure 3.1. Using right-hand rule for determining X, Y, and Z coordinates

The most typical CNC milling machine setup is with the longest axis alongside X coordinate, thus defining the coordinate system can be easily defined applying the right-

24

hand rule, see Figure 3.2. In the case of CNC lathe typically the Z is the longest axis passing through the spindle as shown in Figure 3.3.

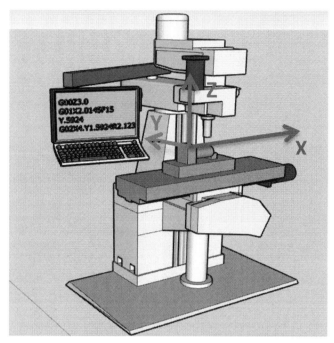

Figure 3.2   3-axis mill. X, Y, Z axis of movement. Z axis through the spindle, X, Y perpendicular to Z and each other

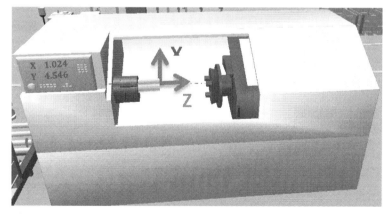

Figure 3.3 CNC lathe with Z and X axis of movement. The primary Z axis pass through the spindle and the secondary X axes is perpendicular to the Z axis

Some machine has more than 2 or 3 moving axes. There are also 4 or 5 axis machines, where the 4 or both 4 and 5 axis are rotational, see Figure 3.4. The direction of the rotational axis can be defined by the right-hand rule for rotation, as shown in Figure 3.5. One can find the positive rotary motion axis by pointing the thumb in the direction of one of the linear axis (X, Y, Z), then the corresponding direction of the rotation axis (A, B, C) can be defined by following the curl of the rest of the fingers.

Figure 3.4 Five-axis CNC milling machine

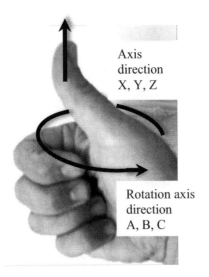

Figure 3.5 Right-hand rule for rotation motion about X, Y, and Z

Turning CNC machine fundamentals

As described before, all CNC turning machines (lathes) use two-axis X-Z coordinate system. The primary axis Z is horizontal alongside the rotation spindle axis, the second axis X is always perpendicular to the primary axis and may have a different configuration (horizontal, vertical or under some angle relative to the base of the machine). Figure 3.6 shows both axes –primary Z and secondary Y. Some CNC lathes may have addition attachment with milling capabilities allowing complex surface machining, in such a case, additional axis linear or rotational are used. Another modern CNC may include dual spindles, allowing machining of both sides of the workpiece without removing it from the machine. The tool turret, which allows automatic changes of tools, can be located in front or behind the Z axis line depending on the configuration of the machines.

The ZX coordinate plane is divided into four quadrants, where the position of each point on the plane can be determined by value and sign of the distance to each axis from the origin.

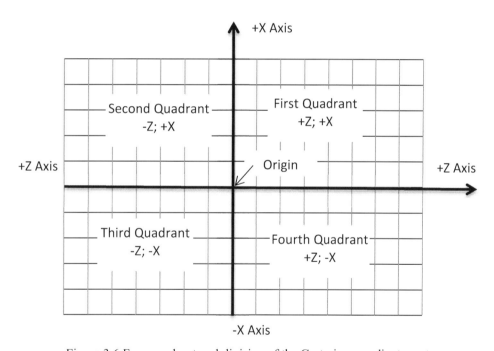

Figure 3.6 Four quadrants subdivision of the Cartesian coordinate system

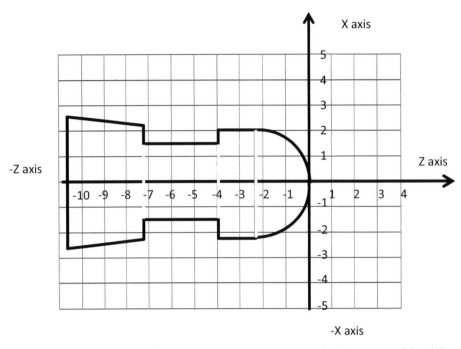

Figure 3.7 Part drawing on Cartesian coordinate system graph. The center of the right front side of the workpiece is usually set up to be at the origin of the coordinate system

The coordinates of each point on profile and their sign the workpiece can be found using the four quadrants coordinate system. Please note that the X values are usually given in diametrical values, see Figures 3.7 and 3.8. For example, if the diameter of a point of the profile of the workpiece is 2.5 inch it will be programmed as X2.5. Since the workpiece profile is symmetrical we can simplify the drawing by showing only the upper half of it on drawings, but we still have to consider diametrical X values.

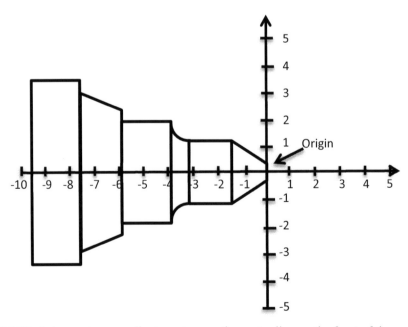

Figure 3.8 Workpiece set on coordinate system on the center line on the front of the workpiece.

28

For convenience, during machining on CNC machine, the origin is usually set on the center line on the front of the workpiece. This is not a rule, just widely accepted practice, even though in some case setting the origin on other feature surface or on the front of the chuck is also used.

Programming in diametrical/radius values

CNC programming allows using diametrical or radius values for X coordinates. Most common programming is diametrical as most of turning parts are design and machined using diametrical measurements. For example, the valise for X on point A is 8.0 since is diametrical value, although in the coordinate system the value is 4.0, as shown in Figure 3.9. To use a correct diametrical value the coordinate X for each point must be doubled.

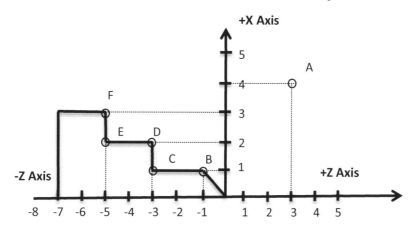

Figure 3.9 Diametrical/radius values

Radius programming is also possible on any CNC machine. In such a case the coordinate value on the graph for X can be used directly to program. For example, point D on the drawing has value 2.0, see Figure 3.9. When programming in radius values caution shall be exercised to calculate correct values, which may be given as diameters on the drawing. When machining again the measure values need to be divided by two to calculate radius values. Rounding errors and tolerance recalculations also may pose problems to achieve the desired accuracy.

CNC machine setting must be changed to set up from diameter to radius value and vice-versa. By default, most CNC machines are set up for diametrical values.

Programming with absolute coordinates

Programming of a point on the profile of the workpiece is measured directly from the origin of the coordinate system. This is so-called absolute programming as the absolute values are measured directly from the origin of the coordinate system. Figure 3.10 shows example absolute coordinate measured from the origin.

How to obtain absolute coordinate values:

For Z measure the distance from a point directly to coordinate axis Z, for X measure the distance from a point directly to coordinate axis X. Remember when using diametrical values to double the X value, but keep the Z value as it is. In addition sign from each value depends on in which quadrant is the measured point. Most users have no difficulties obtaining absolute coordinates for each point, since they had had previous experience from high school/college general classes in mathematics /physics, etc.

Let follow the example shown in Figure 3.10 to find the coordinate values for points A-F.

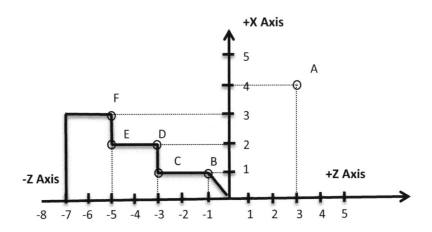

Figure 3.10 Programming with absolute values

The first point is always measured from the origin X0, Z0, of the coordinate system. Each Z-axis point can be found directly below the measured point by traveling directly alongside the Z-axis. Write down the Z value. Next, go up directly to the measured point alongside the X-axis and write down the X value.

EXAMPLE: How to find coordinate values of the point A
1. Start the origin X0Z0
2. Move to the right until you reach exactly bellow point A
3. Go up directly to the measured point A
Result: Diametrical values of the coordinates of point A is X8, Z3
Result: Radial values of the coordinates of point A is X4, Z3

EXAMPLE: How to find coordinate values of the point B
1. Start the origin X0Z0
2. Move to the left until you reach exactly bellow point B
3. Go up directly to the measured point B
Result: Diametrical values of the coordinates of point B is X2, Z-1
Result: Radial values of the coordinates of point B is X1, Z-1

EXAMPLE: How to find coordinate values of the point C
1. Start the origin X0Z0
2. Move to the left until you reach exactly bellow point C

3. Go up directly to the measured point C
Result: Diametrical values of the coordinates of point C is X2, Z-3
Result: Radial values of the coordinates of point C is X1, Z-3

Similarly coordinates of points D, E and F are calculated.

Point D:

Result: Diametrical values of the coordinates of point D is X4, Z-3
Result: Radial values of the coordinates of point D is X2, Z-3

Point E:

Result: Diametrical values of the coordinates of point E is X4, Z-5
Result: Radial values of the coordinates of point E is X2, Z-5

Point F:

Result: Diametrical values of the coordinates of point F is X6, Z-5
Result: Radial values of the coordinates of point E is X3, Z-5

Programming with incremental coordinates

Programming of a point on the profile of the workpiece is measured directly from the previous point, see Figure 3.11. This method is not used very often and may create confusion for some users when calculating the diametrical values.

How to obtain incremental coordinate values:
The first point is measured from the origin of the coordinate system. Each subsequent point is measure from using as a reference the previous point. Remember when using diametrical values to double the X value, but keep the Z value as it is. In addition sign from each value depends not on which quadrant is the measured point but in which direction you measure using the previous point as a reference. Most users have no difficulties obtaining absolute coordinates for each point but may have problems defining the incremental coordinates. One can verify the correctness of the calculated incremental value by finding first the absolute values for each point and then subtract them from the previous reference point. Let follow the example shown in Figure 3.11 to find the incremental coordinate values for points A-F.

31

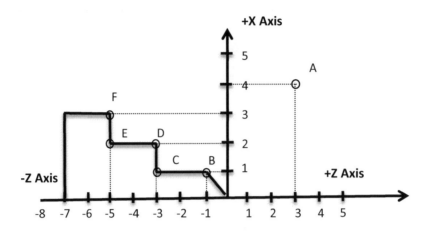

Figure 3.11 Programming with incremental values

Each X/Z coordinate point can be found directly by measuring from the previous point. The first point is always measured from the origin X0, Z0, of the coordinate system.

EXAMPLE: How to find coordinate values of the point A
1. Start the origin X0Z0
2. Move to the right until you reach exactly bellow point A
3. Go up directly to the measured point A
Result: Diametrical values of the coordinates of point A is X8, Z3
Result: Radial values of the coordinates of point A is X4, Z3

EXAMPLE: How to find coordinate values of the point B
1. Start the point A
2. Move to the left until you reach exactly bellow point B
3. Go down directly to the measured point B
Result: Diametrical values of the coordinates of point B is X-6, Z-4
Result: Radial values of the coordinates of point B is X-3, Z-4

EXAMPLE: How to find coordinate values of the point C
1. Start the point B
2. Move to the left until you reach exactly bellow point C
3. Go up directly to the measured point C
Result: Diametrical values of the coordinates of point C is X0, Z-2
Result: Radial values of the coordinates of point C is X0, Z-2

Similarly coordinates of points D, E and F are calculated.
Point D:

Result: Diametrical values of the coordinates of point D is X2, Z0
Result: Radial values of the coordinates of point D is X1, Z0

32

Point E:

Result: Diametrical values of the coordinates of point E is X0, Z-2
Result: Radial values of the coordinates of point E is X0, Z-2

Point F:

Result: Diametrical values of the coordinates of point F is X2, Z0
Result: Radial values of the coordinates of point E is X1, Z0

Tools for Turning

Turning operations are performed using single point tool. Depends on the operation requirements separate tools are needed for rough and finishing, drilling, boring, slotting and cutoff, and threads. It's common in the industry to use tools with indexable inserts. Tools with indexable inserts can be removed rotated, flipped, and reattached without changing the size and shape of the tool. Depends on the manufactures the tool maker produce tools following one of the insert standards the International Organization for Standardization (ISO) and America National Standard Institute (ANSI). Classification based on ISO 1832-1991 and ANSI B212.2.2002 includes: Insert shape, relief angle, tolerances, insert type, size (IC), thickness, corner radius, left or right hand insert, and cutting edge condition. ANSI B212.2.2002 specifies 10 positions (Shape; Clearance; Tolerance class; Type; Size; Thickness; Cutting-point configuration; Edge preparation; Hand; Facet size) denoted by capital letter or number. Each one defines the specific characteristic of the insert as listed below:

1. Shape

Parallelogram	Diamond	Hexagon	Rectangle	Octagon	Pentagon	Round	Square	Triangle	Trigon
A-85°	H-120°	K-120°	L-90°	O-135°	P-108°	R	S-90°	T-60°	W-80°
B-82°	D-55°								
K-55°	E-75°								
	M-86°								
	V-35°								

Some common insert shapes are shown below:

2. Clearance-relief angles
A–3°; B–5°; C–7°; D–15°; E–20°; F–25°; G–30°; N–0°; and P–11°.

3. Tolerance class
There are 14 tolerance classes denoted by letters A, B, C, D, E, F, G, H, J, K, L, M, U and N. For detail refer to the standard ANSI B212.2.2002 or Machinery's Handbook.

4. Type
There are 14 types of inserts with different designs (holes, countersinks, special features, and rakes) denoted by letters A, B, C, D, F, G, H, J, M, N, Q, R, T, U, W, and X. For details refer to the standard ANSI B212.2.2002 or Machinery's Handbook.

5. Size
The size define the inscribe circle (IC) for inserts with Round, Square, Triangle, Trigon, Pentagon, Hexagon, Octagon, and Diamond. One digit for 1/8" (e.g. 1 – 1/8"; 2 – 1/4",

and so on), and two-digits when isn't a whole number (e.g. 1.2 – 5/32"; 1.5 – 3/16" and so on). For details refer to the standard ANSI B212.2.2002 or Machinery's Handbook.

6. Thickness
One or two digit number showing the thickness of the inserts in 1/16". For detail refer to the standard ANSI B212.2.2002 or Machinery's Handbook.

7. Cutting-point configuration
The cutting point configuration can have radius or facet shape. For detail refer to the standard ANSI B212.2.2002 or Machinery's Handbook.

8. Edge preparation
The edge preparation indicated by the capital letter (A, B, C, E, F, J, K, P, S, and T) define the edge treatment and surface finish. For details refer to the standard ANSI B212.2.2002 or Machinery's Handbook.

9. Hand
The hand define type of the tool R-Right hand; L-Left hand; and N-Neutral

10. Facet size
The Facet size is used if there is a letter for Cutting-point configuration in the seventh position. It number represent 1/64". For detail refer to the standard ANSI B212.2.2002 or Machinery's Handbook.

For example, a tool insert with the notation:

1	2	3	4	5	6	7	8	9	10
T	N	M	G	5	4	3			A

Represents a tool with the following parameters:

1-T- Shape is Triangle–60°; 2-N-Relieve angle is 0°; 3-M-Tolerances are: Inscribe circle -0.002-0,004; Thickness-0.005; 4-G-Type is Chip grove both surface with a hole; 5-5-Size-IC size 5 is 5/8"; 6-4-Thickness- is 1/4"; 7-3- Cutting-point configuration-is 3/64". Position 8, 9, and 10 are used when required.

Notes:

Notes:

Notes:

Notes:

Notes:

Chapter 4

CNC
Turning Programming

CNC Turning programming

Letter addresses used in CNC turning

Letter address or word used in CNC turning are followed by a variable value used in programming with G and M codes each state or movement. The most common addresses used in CNC turning are listed below. Refer to operator and program manual for a specific machine for exact words and their usage.

Letter	Description
A	A axis of machine
B	B axis of machine
C	C axis of machine
D	Depth of cut
	Dwell time
	Tool radius compensation number
F	Feed rate
G	Preparatory function
I	X axis center (incremental) for arcs
	Z axis center (incremental) for arcs
	Tread High for G76
M	Miscellaneous function
N	Block number
	Start block canned cycles
	Dwell time in canned cycles and with G4
Q	End block canned cycles
R	Arc radius or canned cycle plane
S	Spindle speed
T	Tool selection

Letter	Description
U	U stock in X direction
	X incremental coordinate
W	W stock in Z direction
	Z incremental coordinate
X	X coordinate
Z	Z coordinate

G and M command used in CNC turning

G and M command used in CNC turning are followed by a variable value used in programming. The most common G and M code command used in CNC turning are listed below. Refer to operator and program manuals for a specific machine for the exact command and their usage.

Code	Parameters	Description	
MOTION	**(X Y Z A B C U V W apply to all motions)**		
G00		Rapid Move	Modal
G01		Linear Interpolation	Modal
G02, G2	I J K or R, P	Circular Interpolation CW	Modal
G03, G3	I J K or R, P	Circular Interpolation CCW	Modal
G04	P	Dwell	
G28	X Z	Automatic Zero Return	
G29	X Z	Return from Zero Return Position	
G32	X Z F	Simple Thread	Modal
CANNED CYCLES	**(X Y Z or U V W apply to canned cycles, depending on active plane)**		
G71	P Q F	Finishing Cycle	
G71	P Q U W D F	Turning Cycle	
G72	P Q U W D F	Facing Cycle	
G76	P Z I J R K Q H L E	Threading Cycle	Modal
DISTANCE MODE			
G90		Absolute Programming	Modal
G91		Incremental Programming	Modal
FEED RATE MODE			
G98		Units per minute feed rate	
G99		Units per revolution	
SPINDLE CONTROL			
M03	S	Spindle Rotation Control - CW	Modal
M04	S	Spindle Rotation Control - CCW	Modal
M05		Stop Spindle	Modal
G96	S D	Constant Surface Speed -CSS	Modal
G97	S D	Constant Spindle Speed Mode (RPM)	Modal
COOLANT			
M08		Coolant Start	Modal
M09		Coolant off	Modal
STOPPING			
M00		Program Pause	

M01		Optional Program Pause	
M02		Program End	
M30		Program End, return to the beginning	
UNITS			
G20		Inch Units	Modal
G21		MM Units	Modal
CUTTER RADIUS COMPENSATION			
G40		Compensation Off	Modal
G41	D	Cutter Compensation Left	Modal
G42	D	Cutter Compensation Right	Modal
RETURN MODE IN CANNED CYCLES			
G98		Canned Cycle Return Level	Modal
OTHER MODAL CODES			
F		Set Feed Rate	Modal
S		Set Spindle Speed	Modal
T		Select Tool	Modal
G54-G59.3		Select Coordinate System	Modal
NON-MODAL CODES			
	T	Tool Change	
G28		Go/Set Predefined Position	
M101 - M199	P Q	User Defined Commands	
COMMENTS & MESSAGES			
/		Block skip	
(...)		Comments	

G00 Rapid Linear Motion

Format structure: G00X__Z__ (G00X__/G00Z__/ G0X__Z__/ G0X__/ G0Z__)

Rapid position the tool to the position specified by the cordites after the command. The rate of movement is the fastest possible for a certain machine. Depending on the machine design it may move directly on straight line to the designated point, see Figure 4.1. On some machines tool moves simultaneously on both axis (at 45° degree angle) and then straight line on the axis direction with remaining value, as show in Figure 4.2. Therefore it is good practice to move only on single axis at one time, thus avoiding possibility of tool collisions with table or workpiece during the rapid movements. Rapid movement to approach point on the workpiece shall be avoided when possible, instead a feed rate motion on desired trajectory shall be used.

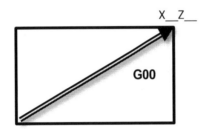

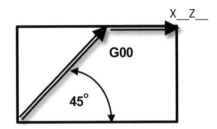

Figure 4.1 Rapid motion directly

Figure 4.2 Rapid motion on 45° degree, rest alongside axis

G0 Command Example:

N50 G0 X3.0 Z1.0 (Rapid move from present coordinates position to X3.0 Z1.0)

N55 Z0.1 (Rapid move from present coordinates position to X3.0 Z0.1)

> *Note: **G0** (modal) use is **optional** and not need to be specified again in the next block **N55**.*

Let make an example program for the part shown in Figure 4.3.

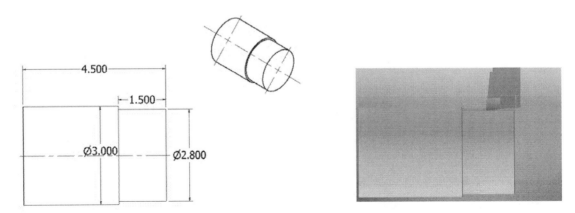

Figure 4.3 G00 Example drawing and tool path

Example program for G00 (Example G00):

Tool: T0101 (tool #1, offset #1), Right hand general turning tool, insert shape C=80° degree
          Diamond, type T

Tool starting position: X3.5, Z3.5.

%	(Start flag)
o1000 (G00)	(Program Number 1000)
N10 G20 G90	(Inch Units, Absolute programming)
N20 T0101	(Tool change Tool#1, offset #1)
N30 M8	(Coolant start)
N40 G97 S500 M3	(RPM speed, speed 500, CW rotation)
N50 G0 X3.0 Z0.1	(Rapid move to X3.0 Z0.1)
N60 G1 Z-3.0 F0.012	(Linear interpolation G1 to Z-3.0, feed rate 0.012)
N70 G1 X3.2	(Linear interpolation G1 to X3.2, federate .012)
N80 G0 Z0.1	(Rapid move to Z0.1)
N90 X2.80	(Rapid move to X2.80)
N100 G1 Z-1.50	(Linear interpolation G1 to Z-1.50)
N110G1 X3.10	(Linear interpolation G1 to X3.10)
N120 G0 Z0	(Rapid move to Z0)
N130 G1 X0	(Linear interpolation G1 to X0)
N140 G1 Z0.1	(Linear interpolation G1 to Z0.1)
N150 G0 X2.0	(Rapid move to X2.0)
N160 Z3.0	(Rapid move to Z3.0)
N170 M9	(Coolant off)
N180 T0100	(Tool#1 offset cancel #00)
N190M5	(Stop spindle)
N200 M30	(Program end)
%	(End flag)

*Note: Feed rate (modal) specified during the first interpolation remain the same
throughout the program unless changed.*

G01 Linear Interpolation

Format structure: G01X__Z__ F__ (G01X__/G01Z__/ G1X__Z__ / G1X__ / G1Z__)

Linear interpolation G01 command executes the movement on a straight line with specified constant feed rate. Cutting of the tool can move simultaneously on both axes X and Z with synchronized feed rate, it also can move along one of the axes X or Z. The G1 is modal, therefore it not need to be specified again if the current mode is G1.

G0 Command Example:

N60 G1 Z0 F0.012   Linear interpolation G1 to X2.3, feed rate 0.012

Let make an example program for the part shown in Figure 4.4.

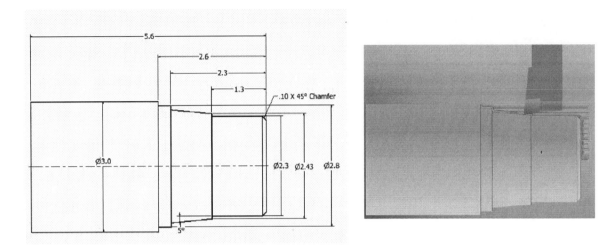

Figure 4.4 G01 Example drawing and tool path

Example program for G01 (Example G01):

Tool: T0202 (tool #2, offset #2), Right hand general turning tool, insert shape C=80°
                         deg., Diamond, type T

Tool starting position: X3.5, Z3.5.

%	(Start flag)
o1001 (G00)	(Program Number 1001)
N10 G20 G90	(Inch Units, Absolute programming)
N20 T0101	(Tool change Tool#1, offset #1)
N30 M8	(Coolant start)
N40 G97 S500 M3	(RPM speed, speed 500, CW rotation)
N50 G0 X2.1 Z0.1	(Rapid move to X2.1.0 Z0.1)
N60 G1 Z0 F0.012	(Linear interpolation G1 to X2.3, feed rate 0.012)
N70 G1 X2.30 Z-0.10	(Linear interpolation X2.30 Z-0.10, same feed rate)
N80 Z-1.4	(Linear interpolation Z-1.3, same feed rate)
N90 X2.43	(Linear interpolation X2.43, same feed rate 0.012)
N100 X2.60 Z-2.3	(Linear interpolation X2.60 Z-2.3 same feed rate)
N110 X2.8	(Linear interpolation X2.8 same feed rate 0.012)
N120 Z-2.6	(Linear interpolation Z-2.6same feed rate 0.012)
N130 X3.0	(Linear interpolation X3.0 same feed rate 0.012)
N140 X3.2	(Linear interpolation X3.2 same feed rate 0.012)
N150 G0 Z0	(Rapid move to Z0)
N160 X2.3	(Rapid move to X2.3)
N170 G1 X0	**(Linear interpolation G1 to X0)**
N180 Z0.1	**(Linear interpolation G1 to Z0.1)**
N190 G0 X2.0	(Rapid move to X2.0)

48

N200 Z3.0	(Rapid move to Z3.0)
N210 M9	(Coolant off)
N220 T0100	(Tool#1 offset cancel #00)
N230 M5	(Stop spindle)
N240 M30	(Program end)
%	(End flag)

Note: Feed rate (modal) specified during the first interpolation remain the same throughout the program unless changed.

G02 Circular Interpolation Clockwise (CW)

Format structure: G02X__Z__F__I__K_ (G2X__Z__F__I__K__)

G02X__Z__F__R__ (G2X__Z__F__R__)

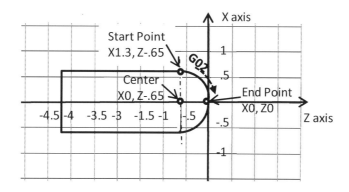

Figure 4.5 G02 Circular Interpolation Clockwise (CW) parameters

Circular interpolation clockwise motion G02 command executes the movement on circular trajectory with specified constant feed rate. Cutting of the tool move simultaneously following circular arc on both axes X and Z with specified radius keeping feed rate synchronized. To calculate the tool motion trajectory on circular art several parameters need to be explicitly defined (Figure 4.5):

1. End point coordinate values X and Z
2. The circular arc radius R or the incremental distance I (in X direction) and K (in Z direction) from the starting point to the center of the arc. Depending on the machine configuration usage of R values may be limited only to 90° or 180° degrees of rotation. There is no limitation if I, and K are used. Note that I, and K are vectors, therefore they have sign +/- defined by the difference in coordinate values between the arc starting point and the center of the arc.
3. Feed rate

The G2 command is modal, therefore there is no need to be specified again if the current mode is G2.

G02 Command Example:

N70 G2 X0 Z0 I-0.65 K0 F.012   90° CW Circular interpolation X0 Z0 I-.65 K0, feed rate .012)

Let make an example program for the part shown in Figure 4.6

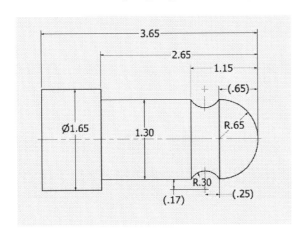 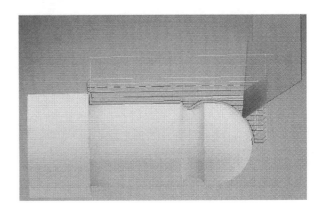

Figure 4.6 G02 example drawing and tool path

Example program for G02 (Example G02):

Tool: T0202 (tool #2, offset #2), Right hand general turning tool, insert shape
                C=55° deg., Diamond, type T
Tool starting position: X3.5, Z3.5.

%	(Start flag)
o1003 (G00)	(Program Number 1003)
N10 G20 G90	(Inch Units, Absolute programming)
N20 T0202	(Tool change Tool#1, offset #1)
N30 M8	(Coolant start)
N40 G97 S1100 M3	(RPM speed, speed 1100, CW rotation)
N50 G0 X1.5 Z-0.65 F.012	(Rapid move to X1.5 Z0.1)
N60 G1 X1.30 F0.012	(Linear interpolation G1.3, feed rate 0.012)
N70 G2 X0 Z0 I-0.65 K0 F.012	(90° CW Circular interpolation X0 Z0 I-.65 K0, feed rate .012)
N80 G1 X1.5	(Linear interpolation Z1.15, feed rate 0.012)
N90 G00 Z-0.65 F.012	(Linear interpolation Z-0.65, feed rate 0.012)
N100 G1 X1.30 F0.012	(Linear interpolation X1.3, feed rate 0.012)
N110 G2 X1.3 Z-1.15 I0.17 K.25 F.012	(Partial CW Circular interpolation X1.30 Z-1.1.5 I0.17 K0.25, feed rate 0.012)
N120 G1 X1.4 F0.15	(Linear interpolation X1.4 same feed rate 0.012)
N130 G0 X2.0	(Rapid move to X2.0)
N140 Z3.0	(Rapid move to Z3.0)

N150 M9	(Coolant off)
N160 T0200	(Tool#1 offset cancel #00)
N170 M5	(Stop spindle)
N180 M30	(Program end)
%	(End flag)

Note: Feed rate (modal) specified during the first interpolation remain the same throughout the program unless changed.

Note that the block N70 and (N110 similar) in the program above can also be defined with arc radius R, instead of I, K values, as shown in the examples below.

G2 Example using radius:

N70 G2 X0 Z0 R0.65 F.012

G03 Circular Interpolation Counter Clockwise (CCW)

Format structure: G03X__Z__F__I__K__ (G3X__Z__F__I__K__)

G03X__Z__ F__R__ (G3X__Z__ F__R__)

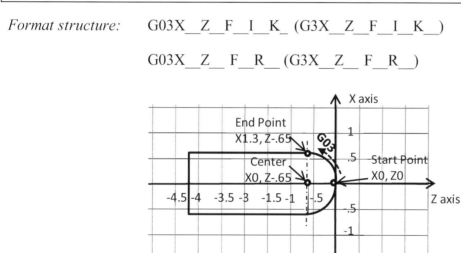

Figure 4.7 G03 Circular Interpolation Counter Clockwise (CCW) parameters

Circular interpolation clockwise motionG03 command executes the movement on circular trajectory with specified constant feed rate. Cutting of the tool move simultaneously following circular arc on both axes X and Z with specified radius keeping feed rate synchronized. To calculate the tool motion trajectory on circular art several parameters need to be explicitly defined (Figure 4.7):

1. End point coordinate values X and Z
2. The circular arc radius R or the incremental distance I (in X direction) and K (in Z direction) from the starting point to the center of the arc. Depending on the machine configuration usage of R values may be limited only to 90° or 180° degrees of rotation. There is no limitation if I, and K are used. Note that I, and K are vectors, therefore they

have sign +/- defined by the difference in coordinate values between the arc starting point and the center of the arc.

3. Feed rate

The G3 command is modal, therefore there is no need to be specified again if the current mode is G3.

G03 Command Example:

N70 G3 X1.30Z-0.65 I0 K-0.65 F.012   90° CCW Circular interpolation X1.30 Z-0.65 I0 K-0.65, feed rate F.015)

Let make an example program for the part shown in Figure 4.8.

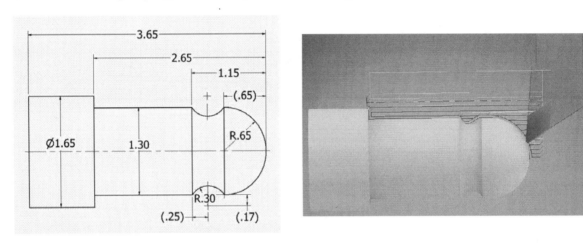

Figure 4.8 G03 example drawing and tool path

Example program for G03 (Example G03):

Tool: T0202 (tool #2, offset #2), Right hand general turning tool, insert shape C=55° deg., Diamond, type T
Tool starting position: X3.5, Z3.5.

	(Start flag)
%	
o1002 (G00)	(Program Number 1002)
N10 G20 G90	(Inch Units, Absolute programming)
N20 T0202	(Tool change Tool#2, offset #2)
N30 M8	(Coolant start)
N40 G97 S1000 M3	(RPM speed, speed 1000, CW rotation)
N50 G0 X1. Z0.1	(Rapid move to X0 Z0.1)
N60 G1 Z0 F0.012	(Linear interpolation G1 to Z0, feed rate 0.012)
N70 G3 X1.30Z-0.65 I0 K-0.65 F.012	(90° CCW Circular interpolation X1.30 Z-0.65 I0 K-0.65, feed rate F.015)
N80 G1 Z1.15	(Linear interpolation Z1.15, same feed rate 0.012)

N90 G3 X1.3 Z-0.65 I0.17 K.25 F.012   (Partial CCW Circular interpolation X1.30 Z-0.65 I0.17
                                                                    K0.25 feed rate F.012)

N100 G1 X1.4 F0.15   (Linear interpolation X1.4 feed rate 0.012)
N110 G0 X2.0   (Rapid move to X2.0)
N120 Z3.0   (Rapid move to Z3.0)
N130 M9   (Coolant off)
N140 T0200   (Tool#1 offset cancel #00)
N150 M5   (Stop spindle)
N160 M30   (Program end)
%   (End flag)

Note: Feed rate (modal) specified during the first interpolation remain the same throughout the program unless changed.

Note that the block N70 (N90 similar) in the program above can also be defined with arc radius R, instead of I, K values, as shown in the examples below.

N70 G3 X1.30Z-0.65 R0.65 F.012

G04 Dwell

Format structure: *G04 P__ (G4 P__)*

         P__ seconds to dwell. P is floating number.

Dwell G04 executes an waiting command in seconds defined by the amount of time specified after P__ . It is frequently used with drilling operation to clear the bottom of the surface of the drilled hole. During the dwell, the feed rate is paused on all axes, while the spindle rotation, coolant, and other operations remain functional, see Figure 4.9. Since the P number is floating, it can be defined in seconds and fraction of a second as well. Note that dwell command is for pause for short time it is not modal and needs to be defined every time when is executed. For longer pause of the program, M00 or M01 shall be used instead.

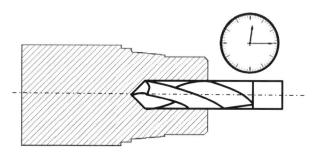

Figure 4.9 G04 Dwell waiting command

G4 Command Example:

N70 G4 P1.05   (Dwell for 1.05 seconds)

The example above show feed rate of tool movements stopped for 1.05 seconds. Let make an example program for the part shown in Figure 4.10.

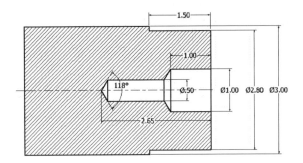 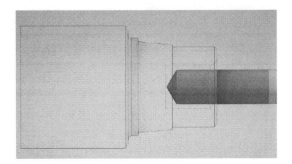

Figure 4.10 G04 Dwell example drawing and tool path

Example program for G04 (ExampleG04):

Tool: **T1515** (tool #15, offset #15), 0.5"Drilling tool,

        T1616 (tool #16, offset #16), 1" Drilling tool.

Tool starting position: X3.5, Z3.5.

%	(Start flag)
o1004 (G00)	(Program Number 1004)
N10 G20 G90	(Inch Units, Absolute programming)
N20 T1515	(Tool change Tool#15, offset #15)
N30 M8	(Coolant start)
N40 G97 S1100 M3	(RPM speed, speed 1100, CW rotation)
N50 G00 X0 Z0.1	(Rapid move to X0 Z0.1)
N60 G1 Z-2.65 F0.012	**(Linear interpolation G1 Z-2.65, feed rate 0.012)**
N70 G4 P1.05	**(Dwell for 1.05 seconds)**
N80 G1 Z0.1	(Rapid move to X3.5 Z3.5)
N90 G00 X3.5 Z3.5 T1500	(Linear interpolation X2.60 Z-2.3 same feed rate)
N100 T1616	(Tool change Tool#16, offset #16)
N110 S800	(RPM speed, speed 600)
N120 G0 X0 Z0.1	(Rapid move to X0 Z0.1)
N130 G1 Z-1.1877 F0.012	**(Linear interpolation Z-1.1877 feed rate 0.012)**
N140 G4 P1.3	**(Dwell for 1.3 seconds)**
N150 G00 X3.5 Z3.5	(Rapid move to X3.5 Z3.5)
N160 T1600	(Tool#16 offset cancel #00)
N170 M9	(Coolant off)
N170 M5	(Stop spindle)
N180 M30	(Program end)
%	(End flag)

Note: Feed rate (modal) specified during the first interpolation remain the same throughout the program unless changed.

G20 Inch Units

Format structure: G20

G20 command set the programming in inch unit. All coordinate values (X, Z) are setup in inches and feed rates in inch per revolution or inch per minute immediately after the issuing of the G20 command. The inch unit command G20 is modal, therefore it not need to be specified again if the current mode is G20. It is practical to setup the G20 at the beginning of the program and not to change units later in the program.

G20 command example:

N10 G20 G90   (Inch Units, Absolute programming)

Let make an example program for the part shown in Figure 4.11

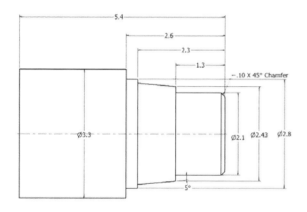 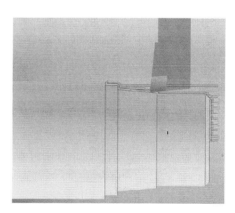

Figure 4.11 G20 example drawing and tool path

Example program for G20 (Example G20):

Tool: T0202 (tool #2, offset #2), Right hand general turning tool, insert shape C=80° deg., Diamond, type T

Tool starting position: X3.5, Z3.5.

```
%                                  (Start flag)
o1001 (G00)                        (Program Number 1001)
N10 G20 G90                        (Inch Units, Absolute programming)
N20 T0101                          (Tool change Tool#1, offset #1)
N30 M8                             (Coolant start)
N40 G97 S500 M3                    (RPM speed, speed 500, CW rotation )
N50 G0 X1.9 Z0.1                   (Rapid move to X2.1.0 Z0.1)
N60 G1 Z0 F0.012                   (Linear interpolation G1 to X2.3, feed rate 0.012)
N70 G1 X2.10 Z-0.10                (Linear interpolation X2.10 Z-0.10, same feed rate)
N80 Z-1.3                          (Linear interpolation Z-1.3, same feed rate 0.012)
N90 X2.43                          (Linear interpolation X2.43, same feed rate 0.012)
N100 X2.60 Z-2.3                   (Linear interpolation X2.60 Z-2.3 same feed rate)
N110 X2.8                          (Linear interpolation X2.8 same feed rate 0.012)
N120 Z-2.6                         (Linear interpolation Z-2.6same feed rate 0.012)
N130 X3.3                          (Linear interpolation X3.0 same feed rate 0.012)
N140 X3.5                          (Linear interpolation X3.2 same feed rate 0.012)
N150 G0 Z0                         (Rapid move to Z0)
N160 X2.3                          (Rapid move to X2.3)
N170 G1 X0                         (Linear interpolation G1 to X0 )
N180 Z0.1                          (Linear interpolation G1 to Z0.1)
N190 G0 X2.0                       (Rapid move to X2.0)
N200 Z3.0                          (Rapid move to Z3.0)
N210 M9                            (Coolant off)
N220 T0100                         (Tool#1 offset cancel #00)
N230 M5                            (Stop spindle)
N240 M30                           (Program end)
%                                  (End flag)
```

Note: Feed rate (modal) specified during the first interpolation remain the same throughout the program unless changed.

G21 Millimeter Units

Format structure: G20

G21 command example:

N10 G21 G90   (Millimeters Units, Absolute programing)

G21 command set the programming in millimeters unit. All coordinate values (X, Z) are setup in millimeters and feed rates in millimeters per revolution or millimeters per minute immediately after the issuing of the G20 command. The millimeters unit command G21 is modal, therefore it not need to be specified again if the current mode is G21. It is practical to setup the G21 at the beginning of the program and not to change units later in the program. Feed rate are also specified in mm per rotation, or mm per minute. Let make an example program for the part shown in Figure 4.12.

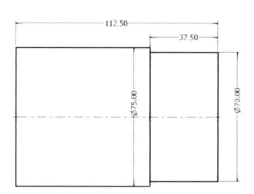 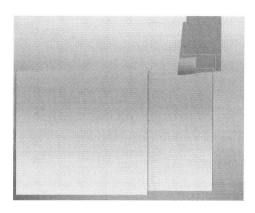

Figure 4.12 G21 example drawing and tool path simulation

Example program for G21 (Example G21):

Tool: T0101 (tool #1, offset #1), Right hand general turning tool, insert shape C=80° degree Diamond, type T

Tool starting position: X87.5, Z87.5.

%	(Start flag)
o1000 (G00)	(Program Number 1000)
N10 G21 G90	(Millimeters Units, Absolute programming)
N20 T0101	(Tool change Tool#1, offset #1)
N30 M8	(Coolant start)
N40 G97 S500 M3	(RPM speed, speed 500, CW rotation)
N50 G0 X75.0 Z0.25	(Rapid move to X75.0 Z0.25)
N60 G1 Z-75.0 F0.3	(Linear interpolation G1 to Z-75.0, feed rate 0.3)
N70 G1 X80.0	(Linear interpolation G1 to X80.0)
N80 G0 Z0.25	(Rapid move to Z0.25)
N90 X70.0	(Rapid move to X70.0)
N100 G1 Z-37.5 F0.3	(Linear interpolation G1 to Z-37.5, feed rate 0.3)
N110 G1 X77.0	(Linear interpolation G1 to X77.0)
N120 G0 Z0	(Rapid move to Z0)
N130 G1 X0	(Linear interpolation G1 to X0)
N140 G1 Z0.25	(Linear interpolation G1 to Z0.25)
N150 G0 X50.0	(Rapid move to X50.0)
N160 Z75.0	(Rapid move to Z75.0)
N170 M9	(Coolant off)
N180 T0100	(Tool#1 offset cancel #00)
N190M5	(Stop spindle)
N200 M30	(Program end)
%	(End flag)

Note: Feed rate (modal) specified during the first interpolation remain the same throughout the program unless changed.

G28 Return to Home Position

Format structure: G28 or G28X_Z

G28 command example:

N130 G28 X4.0 Z3.0   G28 to home position passing through X4.0, Z3.0

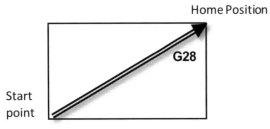

Figure 4.13 Return rapidly directly to Home positon

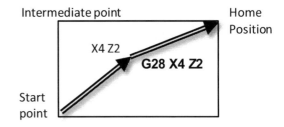

Figure 4.14 Return rapidly to Home positon passing through intermediate position X4 Z2

G28 command return the machine to home position with rapid speed (G00). If there are no coordinate values specified the machine moves rapidly to home (also called machine zero) position, see Figure 4.13. If coordinate values X/ Z are specified the machine moves rapidly to home passing through these coordinate positon, see Figure 4.14. This is used usually when there are obstacle features on the way of the movement of the tool to the home position, therefore it is necessary in such case to define and intermediate position for the tool to pass to avoid collision of the tool on workpiece features.

G29 Return from Home Position

Format structure: G29 or G29X_Z

G29 command example:

N170 G29 X4.0 Z3.0   **(G29 Return to start position passing through** X4.0, Z3.0**)**

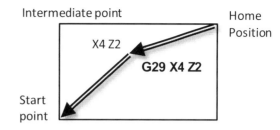

Figure 4.15 Return rapidly directly back from Home positon

Figure 4.16 Return rapidly back from Home positon passing through intermediate position X4 Z2

G29 command is used immediately to return the machine with rapid speed (G00) to starting point, before going to home position after G28, see Figure 4.15. If there are no coordinate values specified the machine moves rapidly to start point. If coordinate values X/Z are specified the machine moves rapidly to the starting point passing through these coordinate positon, see Figure 4.16. This is used usually when there are obstacle features on the way of the movement of the tool from the home position, therefore it is necessary in such a case to define and intermediate position for the tool to pass to avoid collision of the tool on workpiece features. Let make an example program for the part shown in Figure 4.17.

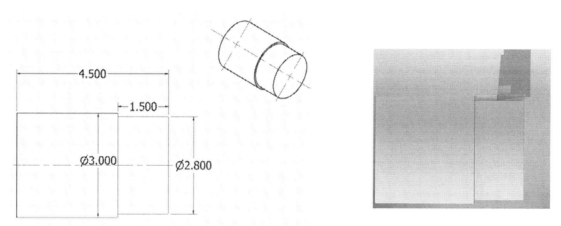

Figure 4.17 G28-G29 example drawing and tool path

Example program for G28-G29 (Example G28-G29):

Tool: T0101 (tool #1, offset #1), Right hand general turning tool, insert shape C=80° degree Diamond, type T

         T0202 (tool #2, offset #2), Right hand general turning tool, insert shape C=55° deg., Diamond, type T

Tool starting position: X3.5, Z3.5.

%   (Start flag)

59

o1028 (G00)	(Program Number 1000)
N10 G20 G90	(Inch Units, Absolute programming)
N20 T0101	(Tool change Tool#1, offset #1)
N30 M8	(Coolant start)
N40 G97 S500 M3	(RPM speed, speed 500, CW rotation)
N50 G0 X3.0 Z0.1	(Rapid move to X3.0 Z0.1)
N60 G1 Z-3.0 F0.012	(Linear interpolation G1 to Z-3.0, feed rate 0.012)
N70 G1 X3.2	(Linear interpolation G1 to X3.2, federate .012)
N80 G0 Z0.1	(Rapid move to Z0.1)
N90 X2.80	(Rapid move to X2.80)
N100 G1 Z-1.50	(Linear interpolation G1 to Z-1.50)
N110G1 X3.10	(Linear interpolation G1 to X3.10)
N120 G0 Z0	**(Rapid move to Z0)**
N130 G28 X4.0 Z3.0	**(G28 to home position passing through X4.0, Z3.0)**
N150 T0100	(Tool#1 offset cancel #00)
N160 T0202	(Tool change Tool#2, offset #2)
N170 G29 X4.0 Z3.0	**(G29 Return to start position passing through X4.0, Z3.0**
	Result: tool moved back to X3.1 Z0)
N180 G1 X0	(Linear interpolation G1 to X0)
N190 G1 Z0.1	(Linear interpolation G1 to Z0.1)
N200 G00 X3.0 Z3.0	(Rapid move to X30, Z3.0)
N210 M09	(Coolant off)
N220 M5	(Stop spindle)
N230 M30	(Program end)
%	(End flag)

Note: Feed rate (modal) specified during the first interpolation remain the same throughout the program unless changed.

G32 Single Thread Cycle

Format structure: G32 X__ Z__ F__

Here - X is the minor diameter, Z is the length, F is the pitch of the thread (pitch=1"/number of threads, for metric the pitch is given with the designation of the thread, e.g. M12x1.5 where M12 is major diameter =12mm, x1.5 is the pitch =1.5mm)

G32 Command example:

N130 G32 Z-2.5 F0.055   Single-pass thread X0.95, Z-2.5, pitch F0.055 (18 threads/inch)

Let make an example program for the part shown in Figure 4.18.

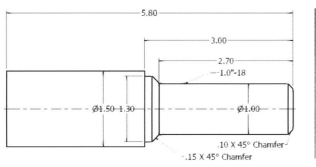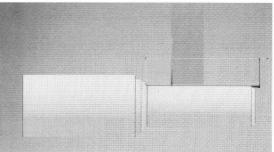

Figure 4.18 G32 example drawing and tool path

G32 is used for simple treading. It provides one pass threading feed synchronized with rotation motion. It can be used for shallow tread or is often the thread output from CAD/CAM software. In former case there are multiple passes calculated where the depth of the thread for each pass is calculated by the computer. Although, it is possible to manually calculate the tool position coordinate values for each point it will be convenient for to use more advance G76 multi-pass threading cycle.

Example program for G32 (Example G32):

Tool: T0101 (tool #1, offset #1), Right hand general turning tool, insert shape C=80° degree Diamond, type T

         T1515 (tool #15, offset #15), Right hand thread tool, V profile C=60° deg[11].

Tool starting position: X3.5, Z3.5.

%	(Start flag)
o1028	(Program Number 1000)
N10 G20 G90	(Inch Units, Absolute programming)
N20 T0101	(Tool change Tool#1, offset #1)
N30 M8	(Coolant start)
N40 G97 S1000 M3	(RPM speed, speed 500, CW rotation)
N50 G0 X1.0 Z0.1	(Rapid move to X3.0 Z0.1)
N60 G1 Z-2.7 F0.012	(Linear interpolation G1 to Z-2.7, feed rate 0.012)
N70 G1 X1.30 Z-2.85	(Linear interpolation G1 to X1.3, Z-2.85, federate .012)
N80 G28 X4.0 Z3.0	(G28 to home position passing through X4.0, Z3.0)
N90 T0100	(Tool#1 offset cancel #00)
N100 T1515	(Tool change Tool#15, offset #15)
N110 G29 X4.0 Z3.0	(G29 Return to start position passing through X4, Z3 Result: tool moved back to X1.3, Z-2.85)
N120 G00 X0.95 Z0.1	(Rapid move to thread start position X1.0, Z0.1)
N130 G32 Z-2.5 F0.055	**(Single-pass thread X0.95, Z-2.5, pitch F0.055 (18 threads/inch)**
N140 G00 X3.0 Z3.0	(Rapid move to X3.0, Z3.0)

N150 M09	(Coolant off)
N160 M5	(Stop spindle)
N170 M30	(Program end)
%	(End flag)

Notes:

1. *Feed rate (modal) specified during the first interpolation remain the same throughout the program unless changed.*

2. *This is sample for a single pass only, for full depth of the thread multiple passes are need to be calculated with X and Z value for each pass.*

G41 Tool Nose Radius Compensation Left

Format structure: G41_

Tool radius compensation is used for the tool nose radius (Figure 4.19) on a single point external or internal (boring) turning tools. It is applied when a profile need to be machined exactly for size and geometry. In many cases when only turning or facing is required it can be omitted because the tool tip setting touches on in Z or X directions.

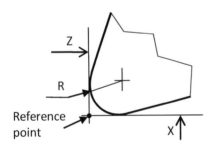

Figure 4.19 Tool nose radius compensation

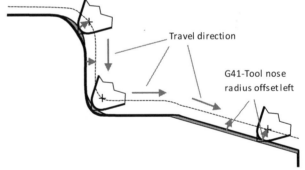

Figure 4.20 Profile errors for tool without radius nose compensation on the left

When machining profile using of a tool the radius on the nose tip will cause profile errors when tool moves simultaneously on both axis. The reference point used for setting of the tool tip is different from the rounded profile of the tool nose radius R, as shown in Figure 4.19. This occur due the fact that the tool tip cut with the rounded part, therefore the error depend the distance from the point of the machining to the radius of the tool as show in Figure 4.20. To avoid such a problem tool nose radius cutter compensation (also called offset) left, relative to the direction of the travel, need to be used as shown in Figure 4.20. Using radius compensation allow programming of the tool path, for turning or boring tools, with actual coordinates without need to recalculate coordinates. The offset value is kept inside CNC controller in Tool Offset Registry

Table. The value imputed during machine setup before. This values is called in the program when tool is called via the tool setup number that is referring to this tool. When the offset is invoked this cause the value stored in setup table to be added or subtracted (multiplied with proper ratio), to the coordinate of tool movement. Tool radius compensation G41 in the program must be followed by actual movement rapid G00 or lineal G01 motion with enough travel distance at least two times more than the tool radius. Tool compensation offset must be canceled with G40 after completion of the tool path to avoid influence the offset value on the next tools or operations. For example T0909 call tool number 9 with its offset number 9, which calls the actual value of the offset from Offset Registry Table.

G41 command example:

N50 G41 G0 X3.6 Z-2.68   **(Tool radius compensation LEFT, Rapid to X3.6 Z-2.68)**

Let make an example program for the part shown in Figure 4.21.

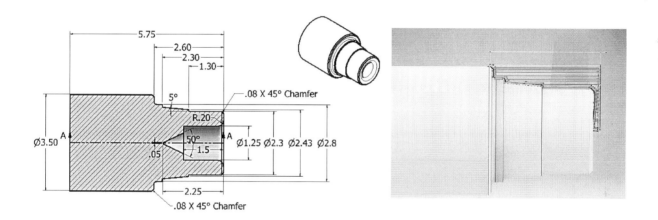

Figure 4.21. G41 example drawing and tool path

Example program for G41 (Example G41):

Tool: T0202 (tool #2, offset #2), Right hand general turning tool, insert shape C=80° deg., Diamond, type T

Tool starting position: X3.5, Z3.5.

%	(Start flag)
o1041	(Program Number 1041)
N10 G20 G90 G40)	(Inch Units, Absolute programming, Cancel Radius Compensation at start of the program)
N20 T0909	(Tool change Tool#9, offset #9)
N30 M8	(Coolant start)
N40 G97 S650 M3	(RPM speed, speed 500, CW rotation)

N50 G41 G0 X3.6 Z-2.68   **(Tool radius compensation LEFT, Rapid to X3.6 Z-2.68)**

N60 G1 X3.5 F0.012   (Linear interpolation X3.5)
N70 G1 X3.34 Z2.60
N80 X2.8
N90 X2.606
N100 X2.43 Z-1.3
N110 X2.3
N120 Z-0.08
N130 X2.14 Z0
N140 X0
N150 X2.3
N160 G1 X0
N170 Z0.1
N180 G40 G0 X3.5.0 Z3.5   (Cancel Radius Compensation, rapid move to X3.5 Z3.5)
N190 M9
N200 T0900   (Tool#9 offset cancel #00)
N210 M5   (Stop spindle)
N220 M30   (Program end)
%   (End flag)

Note: Feed rate (modal) specified during the first interpolation remain the same throughout the program unless changed.

G42 Tool Nose Radius Compensation Right

Format structure: G42_

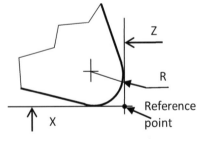

Figure 4.22 Tool nose radius compensation

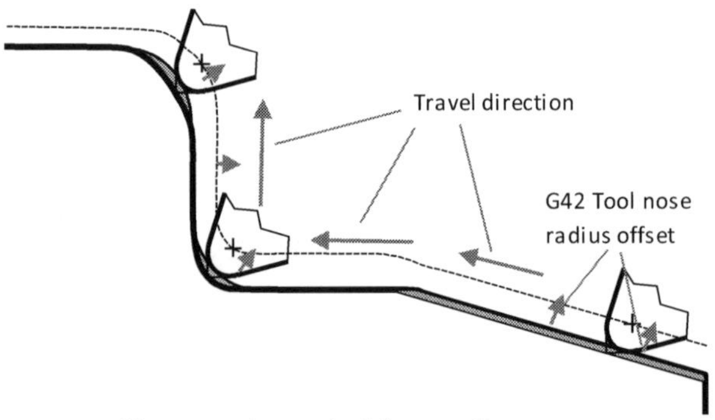

Figure 4.23 Profile errors for tool without radius nose compensation

When machining profile using of the tool nose tip, to calculate tool path, will cause profile errors when tool travel simultaneously on both axis. The reference point used for setting of the tooltip is different from the rounded profile of the tool nose radius R, as shown in Figure 4.22. This occurs due to the fact that the tooltip cut with the rounded part, therefore the error depends the distance from the point of the machining to the radius of the tool as shows in Figure 4.23. To avoid such a problem tool nose radius cutter compensation (also called offset) right, relative to the direction of the travel, need to be used as shown in Figure 4.23. Using radius compensation allow programming of the tool path, for turning or boring tools, with actual coordinates without the need to recalculate coordinates. The offset value is kept inside CNC controller in Tool Offset Registry Table. The value imputed during machine setup before. This value is called in the program when tool is called via the tool setup number that is referring to this tool. When the offset is invoked this cause the value stored in setup table to be added or subtracted (multiplied by proper ratio), to the coordinate of tool movement. Tool radius compensation G41 in the program must be followed by actual movement rapid G00 or lineal G01 motion with enough travel distance at least two times more than the tool radius. Tool compensation offset must be canceled with G40 after completion of the tool path to avoid influence the offset value on the next tools or operations. For example T0606 call tool number 6 with its offset number 6, which calls the actual value of the offset from Offset Registry Table.

G42 command example:

N50 G42 G0 X0 Z0.1   (Tool radius compensation RIGHT, Rapid to X0 Z0.1)

Let make an example program for the part shown in Figure 4.24.

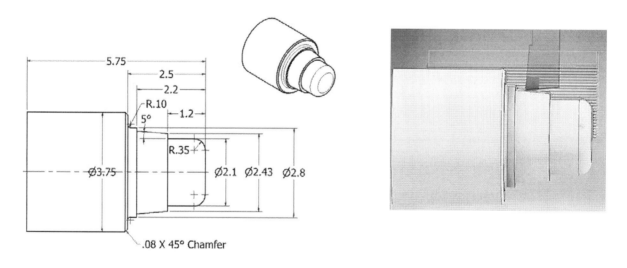

Figure 4.24. G42 example drawing and tool path simulation

Example program for G42 (Example G42):

Tools: T0606 (tool #6, offset #6), Right hand general turning tool, insert shape C=80° deg., Diamond, type T

Tool starting position: X3.5, Z3.5.

%   (Start flag)

o1041	(Program Number 1041)
N10 G20 G90 G40)	(Inch Units, Absolute programming, Cancel Radius Compensation at start of the program)
N20 T0606	(Tool change Tool#6, offset #6)
N30 M8	(Coolant start)
N40 G97 S600 M3	(RPM speed, speed 500, CW rotation)
N50 G42 G0 X0 Z0.1	**(Tool radius compensation RIGHT, Rapid to X0 Z0.1)**
N60 G1 Z0 F0.010	(Linear interpolation Z0, Feed rate 0.010)
N70 G1 X1.4	
N80 G3 X2.1 Z-0.35 R0.35	(or if I, K are used: G3 X2.1 Z-0.35 I0 K-0.35)
N90 G1 Z-1.2	
N100 X2.43	
N110 X2.606 Z-2.2	
N120 X2.8	
N130 Z-2.4	
N140 G2 X3.0 Z-2.5 R0.10	
N150 G1 X3.59	
N150 X3.75 Z-2.66	
N160 G1 X3.8	
N170 G0 Z0.1	
N180 G40 G0 X3.5.0 Z3.5	(Cancel Radius Compensation, rapid move to X3.5 Z3.5)
N190 M9	
N200 T0600	(Tool#6 offset cancel #00)
N210 M5	(Stop spindle)
N220 M30	(Program end)
%	(End flag)

Note: Feed rate (modal) specified during the first interpolation remain the same throughout the program unless changed.

G40 Tool Nose Radius Compensation Cancel

Format structure: G40_

G40 Tool nose radius compensation cancel is used to remove radius compensation invoked with tool nose radius compensation G41-left offset or G42-right offset. Since tool radius compensation is modal, it need to be canceled to avoid confusion. It is also routine G40 to be place at the beginning of the program for remove any remaining radius compensations. Note that Return to Home Position G28 command does not cancel offset, therefore G40 must be used before invoking G28 or any tool changes.

G40 command example:

N180 G40 G0 X3.5.0 Z3.5   **(Cancel Radius Compensation, rapid move to X3.5 Z3.5)**

Let make an example program for the part shown in Figure 4.25.

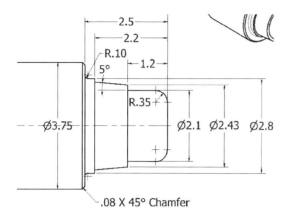 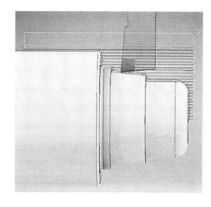

Figure 4.25 G42 example drawing and tool path simulation

Example program for G40 (Example G40):

Tools: T0606 (tool #6, offset #6), Right hand general turning tool, insert shape C=80° deg., Diamond, type T

Tool starting position: X3.5, Z3.5.

%	(Start flag)
o1041	(Program Number 1041)
N10 G20 G90 G40)	**(Inch Units, Absolute programming, Cancel Radius Compensation at start of the program)**
N20 T0606	(Tool change Tool#6, offset #6)
N30 M8	(Coolant start)
N40 G97 S600 M3	(RPM speed, speed 500, CW rotation)
N50 G42 G0 X0 Z0.1	(Tool radius compensation RIGHT, Rapid to X0 Z0.1)
N60 G1 Z0 F0.010	(Linear interpolation Z0, Feed rate 0.010)
N70 G1 X1.4	
N80 G3 X2.1 Z-0.35 R0.35	(or if I, K are used: G3 X2.1 Z-0.35 I0 K-0.35)
N90 G1 Z-1.2	
N100 X2.43	
N110 X2.606 Z-2.2	
N120 X2.8	
N130 Z-2.4	
N140 G2 X3.0 Z-2.5 R0.10	
N150 G1 X3.59	
N150 X3.75 Z-2.66	
N160 G1 X3.8	
N170 G0 Z0.1	
N180 G40 G0 X3.5.0 Z3.5	**(Cancel Radius Compensation, rapid move to X3.5 Z3.5)**
N190 M9	

N200 T0600	(Tool#6 offset cancel #00)
N210 M5	(Stop spindle)
N220 M30	(Program end)
%	(End flag)

Note: Feed rate (modal) specified during the first interpolation remain the same throughout the program unless changed.

G54-G59 Select Work Coordinate System

Format structure: G54_(G55_, G56_, G57_, G58_, G59_)

There are six G-codes (#1-G54, #2-G55, #3-G56, #4-G57, #5-G58, and #6-G59) that can be used to assign workpiece coordinates. They can be used to multi-feature coordinate setup for one or more workpiece, see Figure 4.26. In some CNC machines coordinate can be optionally extended to 48 more using G51.1P1 to P48.

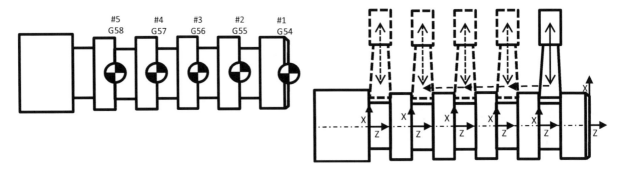

Figure 4.26 G54-59 Commands Example and tool motion

G54-59 command example:

N60 G54   **(Set the coordinate system #1 at position X0 Z0)**

Let make an example program for the part shown in Figure 4.27.

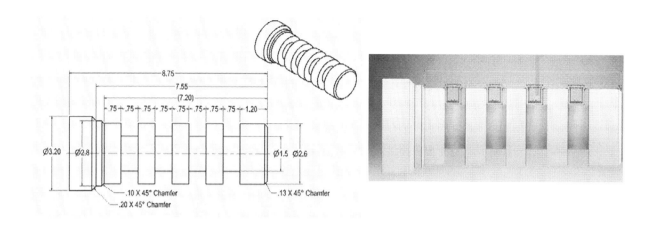

Figure 4.27 G54-59 example drawing and tool path

Example program for G54-G57 (Example G54-G57):

Tool: T1010 (tool #10, offset #10), Slot cutting tool, insert shape C=90° deg.

Tool starting position: X3.5, Z3.5.

%	(Start flag)
o1054	(Program Number 1041)
N10 G20 G90 G40	(Inch Units, Absolute programming, Cancel Radius Compensation at start of the program)
N20 T1010	(Tool change Tool#10, offset #10)
N30 M8	(Coolant start)
N40 G97 S600 M3	(RPM speed, speed 500, CW rotation)
N50 G0 X2.8 Z0.1	(Rapid to X0 Z0.1)
N60 G54	**(Set the coordinate system #1 at position X0 Z0)**
N70 G0 Z-1.95	
N80 G1 X1.5 F0.08	
N90 G1 X2.8	
N100 G55	(Set the coordinate system #2 at position X0 Z-1.95)
N110 G0 Z-1.5	
N120 G1 X1.5 F0.08	
N130 G1 X2.8	
N140 G56	(Set the coordinate system #3 at position X0 Z-3.45)
N150 G0 Z-1.5	
N160 G1 X1.5 F0.08	
N170 G1 X2.8	
N180 G57	(Set the coordinate system #3 at position X0 Z-4.95)
N190 G0 Z-1.5	
N200 G1 X1.5 F0.08	
N210 G1 X2.8	
N220 G54	(Set the coordinate system #1 back to position X0 Z0)
N230 G0 X3.5.0 Z3.5	

69

```
N240 M9
N250 T1000                          (Tool#10 offset cancel #00)
N260 M5                             (Stop spindle)
N270 M30                            (Program end)
%                                   (End flag)
```

Note: 1. Feed rate (modal) specified during the first interpolation remain the same throughout the program unless changed.

2. Program in this example is using tool with the width equal to the size of the channel; in general case, tool width will be smaller requiring multiple cuts.

G70 Finishing Contour Cycle

Format structure: G70 P_ Q_ F_

G70 finishing contour cycle (also called finishing profile cycle) is used after rough cycles such as rough turning cycle G71 or rough facing cycle G72. It is specified by G70 command followed by the starting block letter address P, the finished block letter address Q, and feed rate for finishing F. G70 uses same starting P and finishing blocks address that are specified in rough cycles G71 or G72. It removes the material left over after these finishing cycles following the same contour cycles (defined in G71 or G72). During the machining the contour, the feed rate F specified in G70 block is constant for the entire profile. Please note that the cycle doesn't define the depth of the cut, juts removes the material leftover after rough cycle. The depth of the cut is the difference between the finished contour and rough contour. The amount of material left after the rough cycle is define as offset values U (for X) and W (for Z) specified in rough cycles G71 or G72. G70 can't be specified as standalone command, it shall follow G71 or G72 command. To attain high accuracy it is advisable to use separate tool suitable for the finishing profile G70.

G70 command example:

N210 G70 P70 Q160 F0.007   **(G70 finish profile, P70 start, Q160 end, F feed rate .007)**

Let make an example program for the part shown in Figure 4.28.

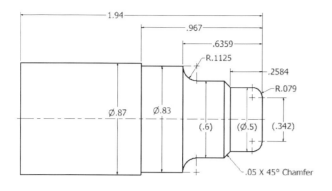 

Figure 4.28 G70 example drawing and tool path simulation

Example program for G70 (Example G70):

Tool: T0404 (tool #4, offset #4) for rough cutting, Right hand general turning tool, insert shape C=80° degree Diamond, type T

                  T0505 (tool #5, offset #5) for finishing cut, Right hand general turning tool, insert shape C=55° deg., Diamond, type T

Tools starting position: X3.5, Z3.5.

%	(Start flag)
o1070	(Program Number 1070)
N10 G20 G90	(Inch Units, Absolute programming)
N20 T0404	(Tool change Tool#4, offset #4 –used for rough profile)
N30 M8	(Coolant start)
N40 G97 S1900 M3	(RPM speed, speed 1900, CW rotation)
N50 G0 X.90 Z0.1	(Rapid to X.9, Z0.1-Starting point of rough cycle)
N60 G71 U.109 R.04	(Rough cut depth U.109, retract after each cut R.07)
N70 G71 P70 Q160 U .04 W .04	(G71rough profile cycle, contour P70 starting, Q 160
F0.014	ending block, F feed rate-0.014 inch per revolution)
N80 G1 X0 Z0	**(First point of the contour - X0, Z0,)**
N90 X.342	
N100 G03 X.5 Z-.079 R.079	
N110 G1 Z-.2584	
N120 X.6 Z-.3084	
N130 Z-.5234	
N140 G2 X.83 Z-.6359 R0.1125	
N150 G1 Z-.967	
N160 X.87	**(Last point of the contour – X.87, Z-.967,)**
N170 G0 X3.5 Z3.5	(rapid move to X3.5 Z3.5)
N180 T0400	(Tool#4 Cancel too offset #00)
N190 T0505	**(change to Tool#5)**

N200 G0 X.9 Z0.1
N210 G70 P70 Q160 F0.007 (G70 finish profile, P70 start, Q160 end, F feed rate .007)

N220 G0 X3.5 Z3.5
N230 T0500 (Tool#5 Cancel too offset #00)
N240 M5 (Stop spindle)
N250 M30 (Program end)
% (End flag)

Note: Feed rate specified during the rough cycle remain the same throughout the profile.

Feed rate specified during the finish cycle remain the same throughout the profile.

G71 Rough Turning Contour Cycle

Format structure: G71 U_ R_   (First block of the rough cycle)

G71 P_ Q_ U _ W_ F_   (Second block of the rough cycle)

G71 rough contour cycle (also called rough profile cycle) is used to remove the excess of material from the stock. G71 can be specified in two sequential blocks, although the first block can be omitted; in such a case the CNC controller uses the default values specified in the machine parameters. In the first block of G71, U defines the depth of the rough cut for each pass and R is the retraction amount after each cut. It the second blocks, G71 command is followed by the starting block letter address P, the finished block letter address Q, U is the stock amount left for finishing in X direction, W is the stock amount left for finishing in Z direction, and F is the feed rate for rough cycle. The program with block numbers following G71, starting from number specifies after P and ending after Q, describe the finished program of the actual contour. G71 removes the material by multiple cutting movements along the Z axis, typically from the front the workpiece to the back side. The depth of each cut is define in the first block (or used the default value from the controller, if the fist block is omitted). Multiple steps allow cutting from cylindrical stock to the profile offset by the U, and W values specified after G71 command. This multiple cutting create a "step" like rough profile that is cleared at the end of the cycle by the tool following the offset profile defined by program with added U and W values. The CNC controller calculate the actual position of starting, finishing, retracting movements based on the parameters specified after G71 command. During the machining of the contour the feed rate F, specified in G71 block, is constant for the entire profile. To attain high accuracy it is advisable to use finishing cycle G70 immediately, after the G71 command, using separate tool.

G71 Command Example:

N60 G71 U.25 R.05 (Rough cut depth U.125, retract after each cut R.05)

N70 G71 P70 Q130 U.04 W.04 F0.016 (G71 rough profile cycle, contour P70 starting, Q 160 ending block, F feed rate-0.016 inch per revolution)

Let make an example program for the part shown in Figure 4.29.

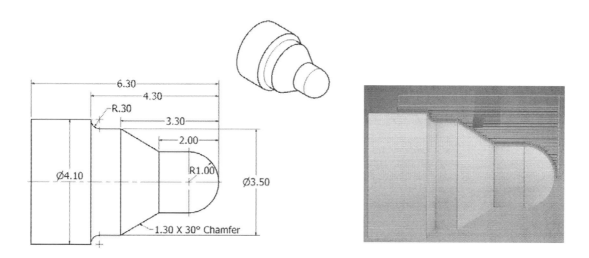

Figure 4.29 G71 example drawing and tool path simulation

Example program for G71 (Example G71):

Tool: T0606 (tool #6, offset #6) for rough cutting, Right hand general turning tool, insert shape C=80° degree Diamond, type T

                   T0707 (tool #7, offset #7) for finishing cut, Right hand general turning tool, insert shape C=55° deg., Diamond, type T

Tools starting position: X3.5, Z3.5.

%	(Start flag)
o1071	(Program Number 1071)
N10 G20 G90	(Inch Units, Absolute programming)
N20 T0606	(Tool change Tool#6, offset #6 –used for rough profile)
N30 M8	(Coolant start)
N40 G97 S850 M3	(RPM speed, speed 1900, CW rotation)
N50 G0 X4.15 Z0.1	(Rapid to X4.15, Z0.1-Starting point of rough cycle)
N60 G71 U.25 R.05	**(Rough cut depth U.125, retract after each cut R.05)**
N70 G71 P80 Q130 U.04 W.04	**(G71rough profile cycle, contour P70 starting, Q 160**
F0.016	**ending block, F feed rate-0.016 inch per revolution)**
N80 G1 X0 Z0	**(First point of the contour - X0, Z0,)**
N90 G03 X2. Z-1.0 R1.0	
N100 G1 Z-2.0	

N110 X3.5 Z-3.30
N120 Z-4.0
N130 G2 X4.1 Z-4.3 R0.3   (Last point of the contour – X.87, Z-.967,)
N140 G0 X3.5 Z3.5   (rapid move to X3.5 Z3.5)
N150 T0600   (Tool#6 Cancel too offset #00)
N160 T0707   (change to Tool#7)
N170 G0 X4.15 Z0.1
N180 G70 P70 Q130 F0.008   (G70 finish profile, P70 start, Q130 end, F feed rate .008)
N190 G0 X3.5 Z3.5
N200 T0700   (Tool#7 Cancel too offset #00)
N210 M5   (Stop spindle)
N220 M30   (Program end)
%   (End flag)

Note: Feed rate specified during the rough cycle remain the same throughout the profile.

Feed rate specified during the finish cycle remain the same throughout the profile.

G72 Rough Facing Contour Cycle

Format structure: G72 U_ R_   (First block of the rough cycle)

                        G72 P_ Q_ U _ W_ F_   (Second block of the rough cycle)

G72 rough contour cycle (also called rough profile cycle) is used to remove the excess of material from the stock. G72 can be specified in two sequential blocks, although the first block can be omitted; in such a case the CNC controller uses the default values specified in the machine parameters. In the first block of G72, U defines the depth of the rough cut for each pass and R is the retraction amount after each cut. It the second blocks, G72 command is followed by the starting block letter address P, the finished block letter address Q, U is the stock amount left for finishing in X direction, W is the stock amount left for finishing in Z direction, and F is the feed rate for rough cycle. The program with block numbers following G72, starting from number specifies after P and ending after Q, describe the finished program of the actual contour. G72 removes the material by multiple cutting movements along the X axis, typically from the back the workpiece to the front side. The depth of each cut is define in the first block (or used the default value from the controller, if the fist block is omitted). Multiple steps allow cutting from cylindrical stock to the profile offset by the U, and W values specified after G72 command. This multiple cutting create a "step" like rough profile that is cleared at the end of the cycle by the tool following the offset profile defined by program with added U and W values. The CNC controller calculate the actual position of starting, finishing, retracting movements based on the parameters specified after G72 command. During the machining of the contour the feed rate F, specified in G72 block, is constant for the entire profile. To attain high accuracy it is advisable to use finishing cycle G70 immediately, after the G72 command, using separate tool.

G72 Command Example:

N60 G72 U.25 R.05   (Rough cut depth U.125, retract after each cut R.05)
N70 G72 P70 Q140 U.04 W.04 F0.012   (G72 rough profile cycle, contour P70 starting, Q140 ending block, F feed rate-0.012 inch per revolution)

Let make an example program for the part shown in Figure 4.30.

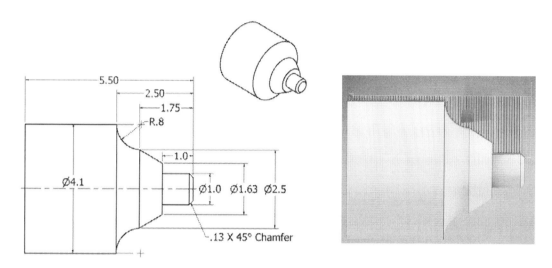

Figure 4.30 G72 example drawing and tool path simulation

Example program for G72 (Example G72):

Tool: T0808 (tool #8, offset #8) for rough cutting, Right hand general turning tool, insert shape C=80° degree Diamond, type T

             T0909 (tool #9, offset #9) for finishing cut, Right hand general turning tool, insert shape C=55° deg., Diamond, type T

Tools starting position: X3.5, Z3.5.

%	(Start flag)
o1072	(Program Number 1072)
N10 G20 G90	(Inch Units, Absolute programming)
N20 T0808	(Tool change Tool#8, offset #8 –used for rough profile)
N30 M8	(Coolant start)
N40 G97 S850 M3	(RPM speed, speed 850, CW rotation)
N50 G0 X4.15 Z0.1	(Rapid to X4.15, Z0.1-Starting point of rough cycle)
N60 G72 U.25 R.05	**(Rough cut depth U.125, retract after each cut R.05)**

```
N70 G72 P80 Q140 U.04 W.04          (G72 rough profile cycle, contour P70 starting, Q140
               F0.012                ending block, F feed rate-0.012 inch per revolution)
N80 G1 X4.10 Z-2.50                 (First point of the contour – X4.1, Z-2.50)
N90 G3 X2.5 Z-1.75 R.8
N100 G1 X1.63 Z-1.0
N110 X1.0
N120 Z-.13
N130 X.74 Z0
N140 X0                            (Last point of the contour – X0, Z0)
N150 G0 X3.5 Z3.5                  (rapid move to X3.5 Z3.5)
N160 T0800                         (Tool#8 Cancel too offset #00)
N170 T0909                         (change to Tool#9)
N180 G0 X4.15 Z0.1
N190 G70 P70 Q130 F0.006           (G70 finish profile, P70 start, Q140 end, F feed rate .006)
N200 G0 X3.5 Z3.5
N210 T0900                         (Tool#9 Cancel too offset #00)
N220 M5                            (Stop spindle)
N230 M30                           (Program end)
%                                  (End flag)
```

Note: Feed rate specified during the rough cycle remain the same throughout the profile.

Feed rate specified during the finish cycle remain the same throughout the profile.

G74 Peck Drilling Cycle

Format structure: *G75 Z__ K__ F__*

Here Z is the position at the bottom of the hole, F is feed rate, and K value is the amount of each peck in Z direction. Note that starting position of the peck drilling cycle is specified in the block before the G74 grove cycle block.

G74 Command Example:

N60 G74 Z-2.65 K.025 F0.012   Peck drilling Cycle, depth of the hole Z-2.65, peck value in
                                   Z direction K.025, feed rate 0.012

Let make an example program for the part shown in Figure 4.31.

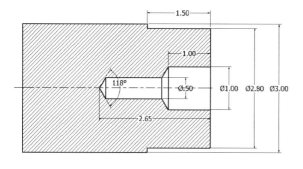 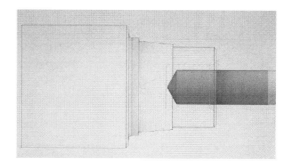

Figure 4.31 G74 Dwell example drawing and tool path

Example program for G74 (ExampleG74):

Tool: **T1515** (tool #15, offset #15), 0.5"Drilling tool,

        T1616 (tool #16, offset #16), 1" Drilling tool.

Tool starting position: X3.5, Z3.5.

%	(Start flag)
o1004 (G00)	(Program Number 1004)
N10 G20 G90	(Inch Units, Absolute programming)
N20 T1515	(Tool change Tool#15, offset #15)
N30 M8	(Coolant start)
N40 G97 S1100 M3	(RPM speed, speed 1100, CW rotation)
N50 G00 X0 Z0.1	(Rapid move to X0 Z0.1)
N60 G74 Z-2.65 K.025 F0.012	**(Peck drilling Cycle, depth of the hole Z-2.65, peck value in Z direction K.025, feed rate 0.012)**
N70 G00 X3.5 Z3.5 T1500	(Linear interpolation X2.60 Z-2.3 same feed rate)
N80 T1616	(Tool change Tool#16, offset #16)
N90 S800	(RPM speed, speed 600)
N100 G74 Z-1.1877 K.02 F0.008	**(Peck drilling Cycle, depth of the hole Z-1.1877, , peck value in Z direction K.02, feed rate 0.008)**
N110 G00 X3.5 Z3.5	(Rapid move to X3.5 Z3.5)
N120 T1600	(Tool#16 offset cancel #00)
N130 M9	(Coolant off)
N140 M5	(Stop spindle)
N150 M30	(Program end)
%	(End flag)

Note: Feed rate (modal) specified during the first interpolation remain the same throughout the program unless changed.

G75 Grove Cycle

Format structure: G75 X__ Z__ F__ I__ K__

Here X value is the diameter of the grove, Z is position of at the end of the grove, F is feed rate, I value is the amount of each peck, and K value is step over amount in Z direction. Note that starting position of the grove cycle is specified in the block before the G75 grove cycle block.

G75 Grove Cycle Command Example:

N130 G75 X.80 Z-.475 F.005 I250 K0   (Grooving cycle X0.80, Z-4.75, feed rate F.005, I250 peck .025, K0 overstep)

Let make an example program for the part shown in Figure 4.32

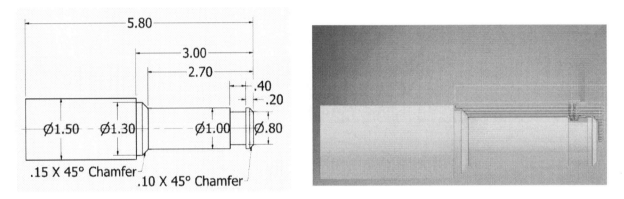

Figure 4.32 G75 example drawing and tool path

Example program for G75 grove cycle (Example G75):

Tool: T0101 (tool #1, offset #1), Right hand general turning tool, insert shape C=80° degree Diamond, type T

T1717 (tool #17, offset #17), Right hand grove tool, square profile, width 1/4" C=90° deg[12].

Tool starting position: X3.5, Z3.5.

```
%                          (Start flag)
o1101                      (Program Number 1101)
N10 G20 G90                (Inch Units, Absolute programming)
N20 T0101                  (Tool change Tool#1, offset #1)
N30 M8                     (Coolant start)
N40 G97 S1000 M3           (RPM speed, speed 500, CW rotation )
N50 G0 X1.0 Z0.1           (Rapid move to X3.0 Z0.1)
N60 G1 Z-2.7 F0.012        (Linear interpolation G1 to Z-2.7, feed rate 0.012)
N70 G1 X1.30 Z-2.85        (Linear interpolation G1 to X1.3, Z-2.85, federate .012)
```

N80 G28 X4.0 Z3.0	(G28 to home position passing through X4.0, Z3.0)
N90 T0100	(Tool#1 offset cancel #00)
N100 T1717	(Tool change Tool#17, offset #17)
N110 G29 X4.0 Z3.0	(G29 Return to start position passing through X4, Z3, Result: tool moved back to X1.3, Z-2.85)
N120 G00 X1.3 Z-0.2	(Rapid move to grove start position X1.3 Z-0.2)
N130 G75 X.80 Z-.475 F.005 I250 **K0**	(Grooving cycle X0.80, Z-4.75, feed rate F.005, I250 peck .025, K overstep)
N140 G00 X3.5	(, Rapid move to X3.5)
N150 Z3.5	(Rapid move to Z3.5)
N160 T1700	(Tool#17 offset cancel #00)
N170 M09	(Coolant off)
N180 M5	(Stop spindle)
N190 M30	(Program end)
%	(End flag)

*Notes:

1. Feed rate (modal) specified during the first interpolation remain the same throughout the program unless changed

---

G76 Threading Cycle

Format structure: G76 X__ Z__ I__ K__ D__ A__ F__ P__

Here X is the thread depth diameter (absolute value), Z is the thread length (absolute value), I is the thread taper amount (radius values), K is the thread depth (radius values), D is the first pass cutting depth, A is the tool nose angle (integer value), F is the feed rate, the lead of the thread, and P is the single edge cutting (load constant). For P value check the operator manual

The G76 threading canned cycle can cut single or multiple straight or tapered threads with different thread profiles, according to the settings. Parameters for the cylindrical there are simplified by using the default vales, see Figure 4.33.

The thread taper amount is specified by I. Thread taper is measured from the target position X, Z at point [7] to position [6]. I value is the difference (radial value) from the start to the end of the thread, it does not represent an angle value.

The thread depth K (or height) is the distance from the crest (top) of the thread to the root (bottom) of the thread. Note that a conventional O.D. taper thread will have a negative I value.

The depth of the first cut through the thread is specified in D. The value for each successive cut is calculated using the equation $D\sqrt{N}$ where N value is the Nth pass along the thread, see Figure 4.34.

The tool nose A angle value (thread angle) can be specified from 0 to 120 degrees depending on the thread type. When A is not specified, the default value 0 degrees is used. To reduce the chatter while threading 60^0 degree included thread it is common to use A59 or A58 for better cutting conditions.

The F code specifies the thread pitch or lead (feed rate for threading). For example for 1/4"-20 thread the pitch is 0.05" (1 inch/20 TPI) and the Lead is 20 TPI. Metric thread specifies the pitch size directly, for example M10x1.25 mans a thread with diameter 10 mm and pitch 1.25 mm.

At the end of the thread an optional chamfer can be performed. The size and angle of the chamfer is controlled by the settings. The chamfer size defined based on the number of threads, so that if 1.000 is specified and the feed rate is .05, then the chamfer will be .05. A chamfer can improve the appearance and functionality of threads that must be machined up to a shoulder. When relief channel is defined at the end of the thread, the chamfer can be eliminated by specifying 0.000 for the chamfer size (Figure4.23.)

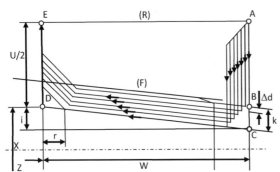

Figure 4.33 G76 Cutting parameters

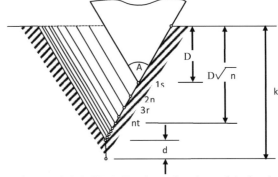

Figure 4.34 G76 Cutting depth and infeed

Thread infeed method defines the way the tool enter into material, in radial directions, at each pass. The first method is radial, also called plunge or perpendicular. It pose cutting disadvantages where both edges of the tool are removing material simultaneously, therefore the load of the tool high and cutting conditions are not good. This produce high heat and tool wear resulting an uneven thread. The second method compound or flank infeed, produces better thread, due to best cutting condition when one side of the tool is removing material, similar to turning feeds machining, therefore the load of the tool and cutting conditions are good. Each of infeed methods have variation (parameter P) to allow better cutting conditions: constant cutting amount, constant cutting depth, one edge cutting, and both edge cutting For details on infeed methods refer to your machine programming manual.

G76 Command Example:

N60 G76 X.4069 Z1.50 I0 K.0328 D.01 A60 F.0769 P2   Thread cycle minor diameter
X.4069, length Z1.50, cylindrical I0, height K.0328, cutting depth D.01, angle A60, pitch F.0769, type P2

Let make an example program for the part shown in Figure 4.35.

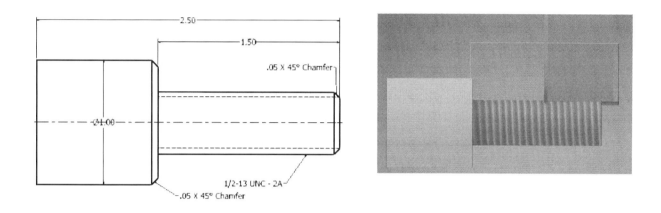

Figure F4.35 G76 example drawing and tool path simulation

Example program for G76 (Example G76):

Tool: T1313 (tool #13, offset #13) for thread cutting, Right hand thread turning tool,
                   insert shape C=60° degree Diamond

Tools starting position: X3.5, Z3.5.

%	(Start flag)
o1076	(Program Number 1076)
N10 G20 G90	(Inch Units, Absolute programming)
N20 T1313	(Tool change Tool#13, offset #13 –thread tool)
N30 M8	(Coolant start)
N40 G97 S800 M3	(RPM speed, speed 800, CW rotation)
N50 G0 X.50 Z0.1	(Rapid to X0.5, Z0.1-Starting point of thread cycle cycle)
N60 G76 X.4069 Z1.50 I0 K.0328	**(Thread cycle minor diameter X.4069, length Z1.50,**
** D.01 A60 F.0769 P2**	**cylindrical I0, height K.0328, cutting depth D.01, angle**
	A60, pitch F.0769, type P2)
N70 G0 X3.5 Z3.5	(rapid move to X3.5 Z3.5)
N80 T1300	(Tool#13 Cancel too offset #00)
N90 M5	(Stop spindle)
N100 M30	(Program end)
%	(End flag)

*Note: Rough and finishing cycle are not show in his program. Thread cycle show for
simplicity.*

Feed rate F specified during the thread cycle G76 is equal to the pitch 1"/number of threads)

G90 Absolute Programming

Format structure: G90

G90 define the coordinate position from the origin of the part coordinate system, specified during the CNC lathe setup, typically setup on the front of the workpiece. All motions of the machine positive (+) or negative (-) are references from the part origin. Absolute programming is the most commonly used in industry, allowing programming without errors since the coordinates are explicitly specified for each motion. This is a modal program and remains active until incremental coordinate G90 is specified. G90 and G91 can be used in the same program.

G90 Command Example:

N10 G90 Absolute programming

Let make an example program for the part shown in Figure 4.36.

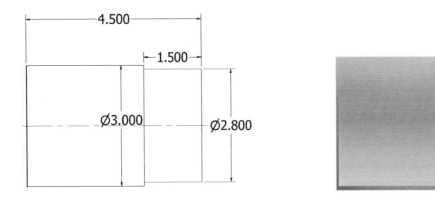

Figure 4.36 G90 example drawing and tool path

Example program for G90 (Example G90):

Tool: T0101 (tool #1, offset #1), Right hand general turning tool, insert shape C=80° degree Diamond, type T

Tool starting position: X3.5, Z3.5.

%	(Start flag)
o1090 (G00)	(Program Number 1090)
N10 G20 G90	**(Inch Units, Absolute programming)**
N20 T0101	(Tool change Tool#1, offset #1)

```
N30 M8                          (Coolant start)
N40 G97 S500 M3                 (RPM speed, speed 500, CW rotation )
N50 G0 X3.0 Z0.1                (Rapid move to X3.0 Z0.1)
N60 G1 Z-3.0 F0.012             (Linear interpolation G1 to Z-3.0, feed rate 0.012)
N70 G1 X3.2                     (Linear interpolation G1 to X3.2, federate .012)
N80 G0 Z0.1                     (Rapid move to Z0.1)
N90 X2.80                       (Rapid move to X2.80)
N100 G1 Z-1.50                  (Linear interpolation G1 to Z-1.50)
N110G1 X3.10                    (Linear interpolation G1 to X3.10)
N120 G0 Z0                      (Rapid move to Z0)
N130 G1 X0                      (Linear interpolation G1 to X0)
N140 G1 Z0.1                    (Linear interpolation G1 to Z0.1)
N150 G0 X2.0                    (Rapid move to X2.0)
N160 Z3.0                       (Rapid move to Z3.0)
N170 M9                         (Coolant off)
N180 T0100                      (Tool#1 offset cancel #00)
N190M5                          (Stop spindle)
N200 M30                        (Program end)
%                               (End flag)
```

Note: Feed rate (modal) specified during the first interpolation remain the same throughout the program unless changed.

G91 Incremental Programming

Format structure: G91

G91 define the incremental position movement from the previous position of the part. The first point of the incremental position is reference from the origin of the coordinate system. All motions of the machine positive (+) or negative (-) are reference from the previous position. Note that when programming in the diametric mode the X values are equal to the difference between the diameters of existing and previous position, not the radial values, also the sign of the incremental values depends on the difference from the previous value. Incremental programming is not often used in industry. This is a modal program and remains active until absolute coordinate G90 is specified. G91 and G90 can be used in the same program multiple times.

G91 Command Example:

N10 G91   Incremental programming

Let make an example program for the part shown in Figure 4.37.

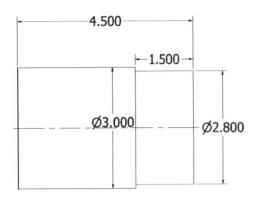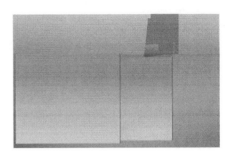

Figure 4.37 G91 example drawing and tool path

Example program for G91 (Example G91):

Tool: T0101 (tool #1, offset #1), Right hand general turning tool, insert shape C=80° degree Diamond, type T

Tool starting position: X3.5, Z3.5.

%	(Start flag)
o1091 (G00)	(Program Number 1091)
N10 G20 G90	**(Inch Units, Absolute programming)**
N20 T0101	(Tool change Tool#1, offset #1)
N30 M8	(Coolant start)
N40 G97 S500 M3	(RPM speed, speed 500, CW rotation)
N50 G0 X3.0 Z0.1	(Rapid move to X3.0 Z0.1)
N60 G91 G1 Z-3.1 F0.012	**(Incremental programming, Linear interpolation G1 to Z-3.0, feed rate 0.012)**
N70 G1 X0.2	(Linear interpolation G1 to X3.2, federate .012)
N80 G0 Z3.1	(Rapid move to Z0.1)
N90 X-.2	(Rapid move to X2.80)
N100 G1 Z-1.6	(Linear interpolation G1 to Z-1.50)
N110G1 X.10	(Linear interpolation G1 to X3.10)
N120 G0 Z1.6	(Rapid move to Z0)
N130 G1 X-2.9	(Linear interpolation G1 to X0)
N140 G1 Z0.1	(Linear interpolation G1 to Z0.1)
N150 G90 G0 X2.0	**(Absolute programming, Rapid move to X2.0)**
N160 Z3.0	(Rapid move to Z3.0)
N170 M9	(Coolant off)
N180 T0100	(Tool#1 offset cancel #00)
N190M5	(Stop spindle)
N200 M30	(Program end)
%	(End flag)

Note: Feed rate (modal) specified during the first interpolation remain the same throughout the program unless changed.

G96 Constant Surface Speed

Format structure: G96

On CNC lathe constant surface speed can be programmed that allow face machining with constant speed when diameter changes. For example when facing cut is done from bigger diameter to smaller one, with a constant speed of rotation, the surface speed changes due to the decrease of the diameter size, resulting in uneven cut and surface finish. To provide better cutting condition G96 constant surface speed can be programmed that can increase or decrease automatically the speed of the rotation, depending on the diameter size, hence keeping the surface speed constant. The surface speed can be increased automatically only up to the maxim limited by the machine capabilities or to the programming maximum rotational speed if it is defined at the beginning of the program by rotating function S.

G96 Command Example:

N40 S1500 M3   RPM speed, **maximum speed 1500**, CW rotation

N130 G96 S470 M3   Constant surface speed at 470 feet/min

Let make an example program for the part shown in Figure 4.38

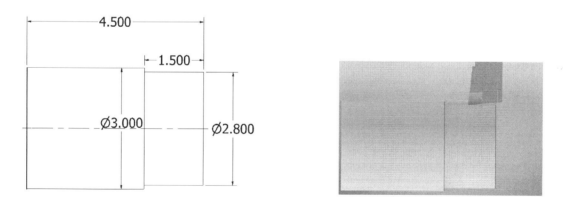

Figure 4.38 G96 example drawing and tool path

Example program for G96 (Example G96):

Tool: T0101 (tool #1, offset #1), Right hand general turning tool, insert shape C=80° degree Diamond, type T

Tool starting position: X3.5, Z3.5.

%   (Start flag)
o1096 (G00)   (Program Number 1096)

```
N10 G20 G90                    (Inch Units, Absolute programming)
N20 T0101                      (Tool change Tool#1, offset #1)
N30 M8                         (Coolant start)
N40 S1500 M3                   (RPM speed, maximum speed 1500, CW rotation )
N50 G0 X3.0 Z0.1               (Rapid move to X3.0 Z0.1)
N60 G1 Z-3.0 F0.012            (Linear interpolation G1 to Z-3.0, feed rate 0.012)
N70 G1 X3.2                    (Linear interpolation G1 to X3.2, federate .012)
N80 G0 Z0.1                    (Rapid move to Z0.1)
N90 X2.80                      (Rapid move to X2.80)
N100 G1 Z-1.50                 (Linear interpolation G1 to Z-1.50)
N110 G1 X3.10                  (Linear interpolation G1 to X3.10)
N120 G0 Z0                     (Rapid move to Z0)
N130 G96 S470 M3               (Constant surface speed at 470 feet/min)
N130 G1 X0                     (Linear interpolation G1 to X0)
N140 G1 Z0.1                   (Linear interpolation G1 to Z0.1)
N150 G0 X2.0                   (Rapid move to X2.0)
N160 Z3.0                      (Rapid move to Z3.0)
N170 M9                        (Coolant off)
N180 T0100                     (Tool#1 offset cancel #00)
N190 M5                        (Stop spindle)
N200 M30                       (Program end)
%                              (End flag)
```

Note: Feed rate (modal) specified during the first interpolation remain the same throughout the program unless changed.

G97 Constant Spindle Speed

Format structure: G97

On CNC lathe constant spindle speed (RPM) can be programmed that keep constant spindle rotation speed when is specified. For example, it can be used after G96 constant speed to return to constant spindle speed (RPM). To provide better cutting condition G97 rotation speed can be change several times within the same program, but the rotation speed will not increase or decrease automatically to keep the surface speed constant as it is with G96. When used in the same program with G96 and thread need to be machined it is important to change to constant spindle speed G97 before the thread cycle started. The reason is that the thread cutting cycle need to synchronize exactly the rotation speed with feed rate to cut a proper thread. Depending on the manufacturer, some CNC controllers will ignore G96 during the thread cutting cycle, while others will stop and show error.

G97 Command Example:

N40 G96 S470 M3   Constant surface speed at 470 feet/min

N150 G97 S1800 M3 Constant spindles speed RPM, speed 1800, CW rotation

Let make an example program for the part shown in Figure 4.39.

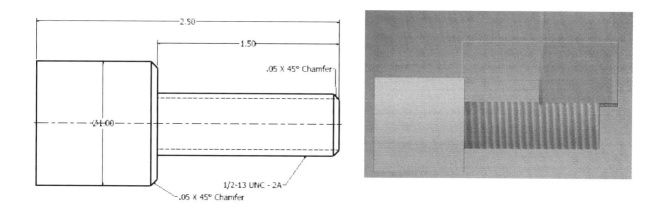

Figure 4.39 G97 example drawing and tool path simulation

Example program for G97 (Example G97):

Tool: T0101 (tool #1, offset #1), Right hand general turning tool, insert shape C=80° degree Diamond, type T

T1313 (tool #13, offset #13) for thread cutting, Right hand thread turning tool, insert shape C=60° degree

Tools starting position: X3.5, Z3.5.

```
%                              (Start flag)
o1097                          (Program Number 1097)
N10 G20 G90                    (Inch Units, Absolute programming)
N20 T0101                      (Tool change Tool#1, offset #1)
N30 M8
N40 G96 S450 M3                (Constant surface speed 470 ft/min, CW rotation )
N50 G00 X1.1 Z.1
N60 G70 P70 Q90 U.01 W.01,
        D1200 F0.012
N70 G1 X0 Z0
N80 X.4
N90 X1. Z-1.505
N100 G70 P70 Q90 F0.06
N110 G00 X.5.Z3.5
N120 T01000
N130 T1313                     (Tool change Tool#13, offset #13 –thread tool)
N140 M8                        (Coolant start)
```

N150 G97 S1800 M3	(Constant spindle speed-1800 RPM, CW rotation)
N160 G0 X.50 Z0.1	(Rapid to X0.5, Z0.1-Starting point of thread cycle cycle)
N170 G76 X.4069 Z1.50 I0 K.0328	(Thread cycle minor diameter X.4069, length Z1.50,
D.01 A60 F.0769 P2	cylindrical I0, height K.0328, cutting depth D.01, angle A60,
	pitch F.0769, type P2)
N180 G0 X3.5 Z3.5	(rapid move to X3.5 Z3.5)
N190 T1300	(Tool#13 Cancel too offset #00)
N200 M5	(Stop spindle)
N210 M30	(Program end)
%	(End flag)

Note: Rough and finishing cycle are not show in his program. Thread cycle show for simplicity.

Feed rate F specified during the thread cycle G76 is equal to the pitch 1"/number of threads)

G98 Feed Rate Per Time

Format structure: G98

G98 command specifies rate per time. Depending on the type of the programming, it can be specified in inch per minute (IPM) or mm per minutes. Since the linear motion is specified as function of time, not RPM, any change in rotation speed will not cause changes of the feed rate that remains constant, while when feed rate per rotation G99 is used the feed rates speed will increase or decrease as it is directly related to the RPM. Note that vales specified in G98 feed rate per minute remain constant until replace by other values or feed rates per rotation G99 is specified. It is common in industry to use feed rate per rotation G99 instead per time G98, since all manual lathes used feed rate per rotation.

G98 Command Example:

N40 G98 S700 M3   Feed rate per inches per minute, RPM speed 700, CW rotation

. . . .

N60 G1 Z0 F2   Linear interpolation Z0, Feed rate F 2 inched per minute

Let make an example program for the part shown in Figure 4.40.

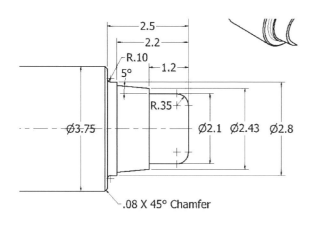 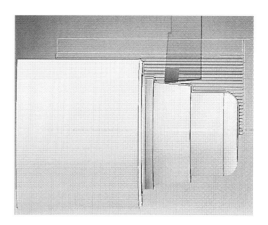

Figure 4.40 G98 example drawing and tool path simulation

Example program for G98 (Example G98):

Tools: T0505 (tool #5, offset #5), Right hand general turning tool, insert shape C=80°
                 deg., Diamond, type T

Tool starting position: X3.5, Z3.5.

%	(Start flag)
o1098	(Program Number 1098)
N10 G20 G90	(Inch Units, Absolute programming)
N20 T0505	(Tool change Tool#5, offset #5)
N30 M8	(Coolant start)
N40 G98 S700 M3	**(Feed rate per inches per minute, RPM speed 700, CW rotation)**
N50 G0 X0 Z0.1	(Tool radius compensation RIGHT, Rapid to X0 Z0.1)
N60 G1 Z0 F2	**(Linear interpolation Z0, Feed rate F 2 inched per minute)**
N70 G1 X1.4	
N80 G3 X2.1 Z-0.35 R0.35	(or if I, K are used: G3 X2.1 Z-0.35 I0 K-0.35)
N90 G1 Z-1.2	
N100 X2.43	
N110 X2.606 Z-2.2	
N120 X2.8	
N130 Z-2.4	
N140 G2 X3.0 Z-2.5 R0.10	
N150 G1 X3.59	
N150 X3.75 Z-2.66	
N160 G1 X3.8	
N170 G0 Z0.1	
N180 G0 X3.5.0 Z3.5	(Rapid move to X3.5 Z3.5)
N190 M9	
N200 T0500	(Tool#5 offset cancel #00)

N210 M5   (Stop spindle)
N220 M30   (Program end)
%   (End flag)

*Note: Feed rate (modal) specified during the first interpolation (F2 feed per minute)
remain the same throughout the program unless changed.*

G99 Feed Rate Per Revolution

Format structure: G99

G99 command specifies rate per revolution. Depending on the type of the programming, it can be specified in inch per rotation (IPR) or mm per rotation. Since the linear motion is specified as a function of rotation any change in rotation speed will cause changes in the feed rate when feed rate per rotation is used the feed rates speed will increase or decrease as it is directly related to the RPM. Note that the values specified in G99 feed rate per minute remain constant until replaced by other values or feed rates per minutes G98 is specified. It is common in the industry to use feed rate per rotation G98, since all manual lathes used feed rate per rotation.

G98 Command Example:

N40 G99 S900 M3   Feed rate per inches per rotation, RPM speed 900, CW rotation

.....

N60 G1 Z0 F.014   Linear interpolation Z0, Feed rate F .014 inched per revolution

Let make an example program for the part shown in Figure 4.41.

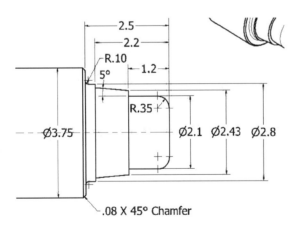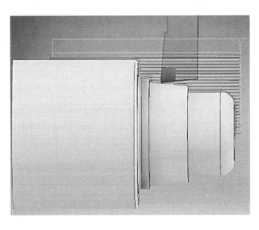

Figure 4.41 G99 example drawing and tool path simulation

Example program for G99 (Example G99):

Tools: T0707 (tool #7, offset #7), Right hand general turning tool, insert shape C=80° deg., Diamond, type T

Tool starting position: X3.5, Z3.5.

%	(Start flag)
o1098	(Program Number 1098)
N10 G20 G90	(Inch Units, Absolute programming)
N20 T0505	(Tool change Tool#5, offset #5)
N30 M8	(Coolant start)
N40 G99 S700 M3	**(Feed rate per inches per minute, RPM speed 700, CW rotation)**
N50 G0 X0 Z0.1	(Tool radius compensation RIGHT, Rapid to X0 Z0.1)
N60 G1 Z0 F.014	**(Linear interpolation Z0, Feed rate F0.14 inched per revolution)**
N70 G1 X1.4	
N80 G3 X2.1 Z-0.35 R0.35	(or if I, K are used: G3 X2.1 Z-0.35 I0 K-0.35)
N90 G1 Z-1.2	
N100 X2.43	
N110 X2.606 Z-2.2	
N120 X2.8	
N130 Z-2.4	
N140 G2 X3.0 Z-2.5 R0.10	
N150 G1 X3.59	
N150 X3.75 Z-2.66	
N160 G1 X3.8	
N170 G0 Z0.1	
N180 G0 X3.5.0 Z3.5	(Rapid move to X3.5 Z3.5)
N190 M9	
N200 T0500	(Tool#5 offset cancel #00)
N210 M5	(Stop spindle)
N220 M30	(Program end)
%	(End flag)

Note: Feed rate (modal) specified during the first interpolation (F.014 feed per revolution) remain the same throughout the program unless changed.

M00 Program Stop

Format structure: M00

M00 Program stop is used when temporary stop needs to be performed during the CNC machining process. When M00 command block is reached the CNC controller stops temporary all essential functions like the spindle rotation, coolant, all axis motions, etc. M00 does not terminate the program execution but is used to stop temporary the program execution so the operator can do some check or adjustment on the machine like coolant hose direction, tool wear or breakage, machined surface size and finish, remove chips from working zone and others. The program is resumed normally when cycle start is pressed by the operator. All setting and order of the operations including coordinate positions, feed rate, speed, etc., remain unchanged after resuming the program execution.

M00 Command Example:

N110 M00   **Program Stop**

The example above shows all movements stopped. Let make an example program for the part shown in Figure 4.42

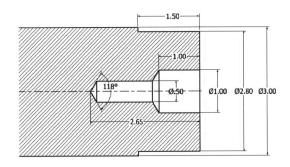 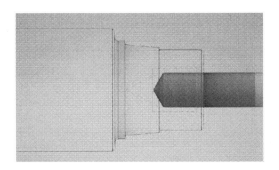

Figure 4.42 M00 Program stop drawing and tool path

Example program for M00 (Example M00):

Tool: **T1212** (tool #12, offset #12), 0.5"Drilling tool,

          T1414 (tool #14, offset #14), 1" Drilling tool.

Tool starting position: X3.5, Z3.5.

%	(Start flag)
o1100 (G00)	(Program Number 1100)
N10 G20 G90	(Inch Units, Absolute programming)
N20 T1212	(Tool change Tool#12, offset #12)
N30 M8	(Coolant start)
N40 G97 S1100 M3	(RPM speed, speed 1100, CW rotation)
N50 G00 X0 Z0.1	(Rapid move to X0 Z0.1)
N60 G1 Z-2.65 F0.012	(Linear interpolation G1 Z-2.65, feed rate 0.012)
N70 G4 P1.05	(Dwell for 1.05 seconds)
N80 G1 Z0.1	(Rapid move to X3.5 Z3.5)
N90 G00 X3.5 Z3.5 T1200	(Linear interpolation X2.60 Z-2.3 same feed rate)
N100 T1414	(Tool change Tool#12, offset #14)
N110 M00	**(Program stop)**

```
N120 S800                    (RPM speed, speed 600)
N130 G0 X0 Z0.1              (Rapid move to X0 Z0.1)
N140 G1 Z-1.1877 F0.012      (Linear interpolation Z-1.1877 feed rate 0.012)
N150 G4 P1.3                 (Dwell for 1.3 seconds)
N160 G00 X3.5 Z3.5           (Rapid move to X3.5 Z3.5)
N170 T1400                   (Tool#14 offset cancel #00)
N180 M9                      (Coolant off)
N190 M5                      (Stop spindle)
N200 M30                     (Program end)
%                            (End flag)
```

Note: Feed rate (modal) specified during the first interpolation remain the same throughout the program unless changed.

M01 Optional Program Stop

Format structure: M01

M01 Optional Program stop is used when temporary stop needs to be performed during the CNC machining process. When M01 command block is reached the CNC controller stops temporary all essential functions like the spindle rotation, coolant, all axis motions, etc. M00 does not terminate the program execution but is used to stop temporary the program execution so the operator can do some check or adjustment on the machine like coolant hose direction, tool wear or breakage, machined surface size and finish, remove chips from working zone and others. The program is resumed normally when cycle start is pressed by the operator. All setting and order of the operations including coordinate positions, feed rate, speed, etc., remain unchanged after resuming the program execution.

M01 works the same way as M00 program stop when an optional M01 switch on the control panel is activated. When the optional button M01 is not active the M01 optional stop is ignored by the controller and program executes normally without any stop function. The M01 optional stop is useful when an initial program is created and refined, allowing troubleshooting program errors, and restoring the normal execution cycle later when the problem is resolved.

M01 Command Example:

N110 M01   Optional program stop

...

N160 M01   Optional program stop

The example above shows all movements stopped. Let make an example program for the part shown in Figure 4.43

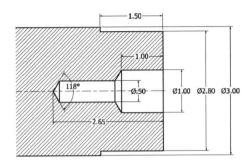 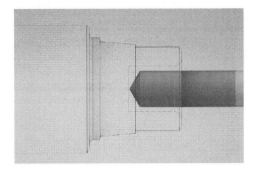

Figure 4.43 M01 Optional program stop example drawing and tool path

Example program for M00 (Example M00):

Tool: **T1212** (tool #12, offset #12), 0.5"Drilling tool,

           T1414 (tool #14, offset #14), 1" Drilling tool.

Tool starting position: X3.5, Z3.5.

%	(Start flag)
o1100 (G00)	(Program Number 1100)
N10 G20 G90	(Inch Units, Absolute programming)
N20 T1212	(Tool change Tool#12, offset #12)
N30 M8	(Coolant start)
N40 G97 S1100 M3	(RPM speed, speed 1100, CW rotation)
N50 G00 X0 Z0.1	(Rapid move to X0 Z0.1)
N60 G1 Z-2.65 F0.012	(Linear interpolation G1 Z-2.65, feed rate 0.012)
N70 G4 P1.05	(Dwell for 1.05 seconds)
N80 G1 Z0.1	(Rapid move to X3.5 Z3.5)
N90 G00 X3.5 Z3.5 T1200	(Linear interpolation X2.60 Z-2.3 same feed rate)
N100 T1414	(Tool change Tool#14, offset #14)
N110 M01	**(Program optional stop)**
N120 S800	(RPM speed, speed 600)
N130 G0 X0 Z0.1	(Rapid move to X0 Z0.1)
N140 G1 Z-1.1877 F0.012	(Linear interpolation Z-1.1877 feed rate 0.012)
N150 G4 P1.3	(Dwell for 1.3 seconds)
N160 M01	**(Program optional stop)**
N150 G00 X3.5 Z3.5	(Rapid move to X3.5 Z3.5)
N160 T1400	(Tool#14 offset cancel #00)
N170 M9	(Coolant off)
N170 M5	(Stop spindle)
N180 M30	(Program end)

%   (End flag)

Note: Feed rate (modal) specified during the first interpolation remain the same throughout the program unless changed.

M02 Program End

Format structure: M02

M02 defines the end of the program execution, terminated the program. M02 is the last block of the CNC programs. After M02 the CNC controlled switches off all operations – spindle rotation all axis feed fate and rapid movements, etc.

M02 Command Example:

N240 M02   Program end

Let make an example program for the part shown in Figure 4.44.

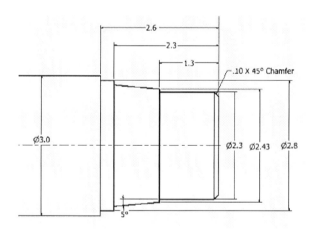 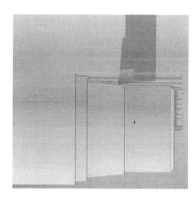

Figure 4.44 M02 example drawing and tool path

Example program for M02 (Example M02):

Tool: T0202 (tool #2, offset #2), Right hand general turning tool, insert shape C=80° deg., Diamond, type T

                  Tool starting position: X3.5, Z3.5.

%   (Start flag)
o1102 (G00)   (Program Number 1102)
N10 G20 G90   (Inch Units, Absolute programming)

N20 T0101	(Tool change Tool#1, offset #1)
N30 M8	(Coolant start)
N40 G97 S500 M3	(RPM speed, speed 500, CW rotation)
N50 G0 X2.1 Z0.1	(Rapid move to X2.1.0 Z0.1)
N60 G1 Z0 F0.012	(Linear interpolation G1 to X2.3, feed rate 0.012)
N70 G1 X2.30 Z-0.10	(Linear interpolation X2.30 Z-0.10, same feed rate)
N80 Z-1.4	(Linear interpolation Z-1.3, same feed rate)
N90 X2.43	(Linear interpolation X2.43, same feed rate 0.012)
N100 X2.60 Z-2.3	(Linear interpolation X2.60 Z-2.3 same feed rate)
N110 X2.8	(Linear interpolation X2.8 same feed rate 0.012)
N120 Z-2.6	(Linear interpolation Z-2.6same feed rate 0.012)
N130 X3.0	(Linear interpolation X3.0 same feed rate 0.012)
N140 X3.2	(Linear interpolation X3.2 same feed rate 0.012)
N150 G0 Z0	(Rapid move to Z0)
N160 X2.3	(Rapid move to X2.3)
N170 G1 X0	(Linear interpolation G1 to X0)
N180 Z0.1	(Linear interpolation G1 to Z0.1)
N190 G0 X2.0	(Rapid move to X2.0)
N200 Z3.0	(Rapid move to Z3.0)
N210 M9	(Coolant off)
N220 T0100	(Tool#1 offset cancel #00)
N230 M5	(Stop spindle)
N240 M02	**(Program end)**
%	(End flag)

Note: Feed rate (modal) specified during the first interpolation remain the same throughout the program unless changed.

M03 Spindle Clockwise Rotation (CW)

Format structure: M03

M03 defines the spindle clockwise rotation (CW), it is usually issued together with spindle speed rotation S defines by a number following it (e.g. S800). M3 is active until the spindle is stopped by M05 command or program stopped. Most of the CNC lathes don't stop the spindle when the tool is changed. If the spindle it stopped in a program, M03 will reactive the rotation with the same rotation speed defined before stopping of the spindle.

M3 Command Example:

N40 G97 S1000 M3   **RPM speed 1000, CW rotation**

Let make an example program for the part shown in Figure 4.45

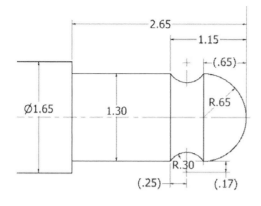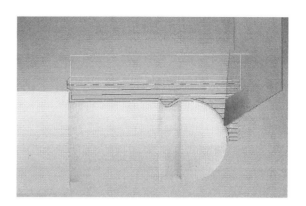

Figure 4.45 M3 example drawing and tool path

Example program for M03 (Example M03):

Tool: T0202 (tool #2, offset #2), Right hand general turning tool, insert shape C=55° deg., Diamond, type T

Tool starting position: X3.5, Z3.5.

%	(Start flag)
o1002 (G00)	(Program Number 1002)
N10 G20 G90	(Inch Units, Absolute programming)
N20 T0202	(Tool change Tool#2, offset #2)
N30 M8	(Coolant start)
N40 G97 S1000 M3	**(RPM speed, speed 1000, CW rotation)**
N50 G0 X1. Z0.1	(Rapid move to X0 Z0.1)
N60 G1 Z0 F0.012	(Linear interpolation G1 to Z0, feed rate 0.012)
N70 G3 X1.30Z-0.65 I0 K-0.65 F.012	(90° CCW Circular interpolation X1.30 Z-0.65 I0 K-0.65, feed rate F.015)
N80 G1 Z1.15	(Linear interpolation Z1.15, same feed rate 0.012)
N90 G3 X1.3 Z-0.65 I0.17 K.25 F.012	(Partial CCW Circular interpolation X1.30 Z-0.65 I0.17 K0.25 feed rate F.012)
N100 G1 X1.4 F0.15	(Linear interpolation X1.4 feed rate 0.012)
N110 G0 X2.0	(Rapid move to X2.0)
N120 Z3.0	(Rapid move to Z3.0)
N130 M9	(Coolant off)
N140 T0200	(Tool#1 offset cancel #00)
N150 M5	**(Stop spindle)**
N160 M30	(Program end)
%	(End flag)

Note: Feed rate (modal) specified during the first interpolation remain the same throughout the program unless changed.

M04 Spindle Counter Clockwise Rotation (CCW)

Format structure: M04

M04 defines the spindle counterclockwise rotation (CCW), it is usually issued together with spindle speed rotation S defines by a number following it (e.g. S900). M4 is active until the spindle is stopped by M05 command or program stopped. Most of the CNC lathes don't stop the spindle when the tool is changed. If the spindle it stopped in a program, M04 will reactive the rotation with the same rotation speed defined before stopping of the spindle.

M4 Command Example:

N40 S1300 M4   **maximum speed 1300**, CCW rotation

Let make an example program for the part shown in Figure 4.46.

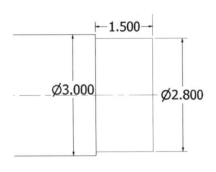
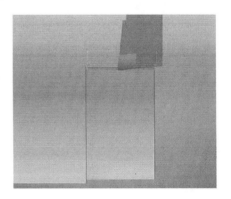

Figure 4.46 M4 example drawing and tool path

Example program for M4 (Example M4):

Tool: T0101 (tool #1, offset #1), Right hand general turning tool, insert shape C=80° degree Diamond, type T

Tool starting position: X3.5, Z3.5.

%	(Start flag)
o1104 (G00)	(Program Number 1104)
N10 G20 G90	(Inch Units, Absolute programming)
N20 T0101	(Tool change Tool#1, offset #1)
N30 M8	(Coolant start)
N40 S1300 M4	(RPM speed, **maximum speed 1300**, CCW rotation)
N50 G0 X3.0 Z0.1	(Rapid move to X3.0 Z0.1)
N60 G1 Z-3.0 F0.012	(Linear interpolation G1 to Z-3.0, feed rate 0.012)
N70 G1 X3.2	(Linear interpolation G1 to X3.2, federate .012)

N80 G0 Z0.1	(Rapid move to Z0.1)
N90 X2.80	(Rapid move to X2.80)
N100 G1 Z-1.50	(Linear interpolation G1 to Z-1.50)
N110 G1 X3.10	(Linear interpolation G1 to X3.10)
N120 G0 Z0	(Rapid move to Z0)
N130 G96 S470 M3	(Constant surface speed at 470 feet/min)
N130 G1 X0	(Linear interpolation G1 to X0)
N140 G1 Z0.1	(Linear interpolation G1 to Z0.1)
N150 G0 X2.0	(Rapid move to X2.0)
N160 Z3.0	(Rapid move to Z3.0)
N170 M9	(Coolant off)
N180 T0100	(Tool#1 offset cancel #00)
N190 M5	(Stop spindle)
N200 M30	(Program end)
%	(End flag)

Note: Feed rate (modal) specified during the first interpolation remain the same throughout the program unless changed.

M05 Spindle Stop

Format structure: M05

M05 stops the spindle rotation. It is used to stop the spindle rotation of the end of the program permanently. Other Mo command like M00, M01, and M02 stop the spindle temporary allowing to restart the rotation.

M5 Command Example:

N170 M5 **Stop spindle**

Let make an example program for the part shown in Figure 4.47.

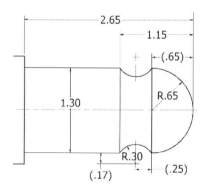 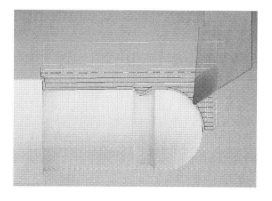

Figure 4.47 M5 example drawing and tool path

Example program for M05 (Example M05):

Tool: T0202 (tool #2, offset #2), Right hand general turning tool, insert shape
C=55° deg., Diamond, type T

Tool starting position: X3.5, Z3.5.

%	(Start flag)
o1105 (G00)	(Program Number 1105)
N10 G20 G90	(Inch Units, Absolute programming)
N20 T0202	(Tool change Tool#1, offset #1)
N30 M8	(Coolant start)
N40 G97 S1100 M3	(RPM speed, speed 1100, CW rotation)
N50 G0 X1.5 Z-0.65 F.012	(Rapid move to X1.5 Z0.1)
N60 G1 X1.30 F0.012	(Linear interpolation G1.3, feed rate 0.012)
N70 G2 X0 Z0 I-0.65 K0 F.012	(90° CW Circular interpolation X0 Z0 I-.65 K0, feed rate .012)
N80 G1 X1.5	(Linear interpolation Z1.15, feed rate 0.012)
N90 G00 Z-0.65 F.012	(Linear interpolation Z-0.65, feed rate 0.012)
N100 G1 X1.30 F0.012	(Linear interpolation X1.3, feed rate 0.012)
N110 G2 X1.3 Z-1.15 I0.17 K.25 F.012	(Partial CW Circular interpolation X1.30 Z-1.1.5 I0.17 K0.25, feed rate 0.012)
N120 G1 X1.4 F0.15	(Linear interpolation X1.4 same feed rate 0.012)
N130 G0 X2.0	(Rapid move to X2.0)
N140 Z3.0	(Rapid move to Z3.0)
N150 M9	(Coolant off)
N160 T0200	(Tool#1 offset cancel #00)
N170 M5	**(Stop spindle)**
N180 M30	(Program end)
%	(End flag)

Note: Feed rate (modal) specified during the first interpolation remain the same throughout the program unless changed.

M08 Coolant Start

Format structure: M08

M08 command starts the cutting fluid (coolant) flow. It is advisable to start the cutting fluid just before first cutting occurred. Cutting fluid is essential in cutting most of the metal alloys, it cools the workpiece and tool material, provides lubrication, and remove chips away from the working area.

M08 Command Example:

N30 M08   Coolant start

Format structure: M09

M09 command switches off the cutting fluid (coolant) flow. It is advisable to stop the cutting fluid just after the last cutting occurred. Depending on the CNC controller the coolant is automatically switched off before/after the tool change, at program stop or end. Refer to operating manual for details regarding starting or stopping of the cutting fluid flow within the program.

M09 Command Example:

N130 M9   Coolant off

Let make an example program for the part shown in Figure 4.48.

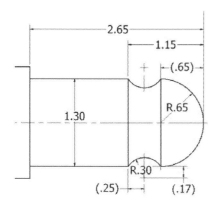
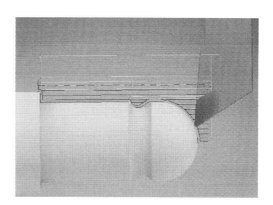

Figure 4.48 M08-M09 example drawing and tool path

Example program for M08-M09 (Example M08-M09):

Tool: T0202 (tool #2, offset #2), Right hand general turning tool, insert shape C=55° deg., Diamond, type T

Tool starting position: X3.5, Z3.5.

%	(Start flag)
o1002 (G00)	(Program Number 1002)
N10 G20 G90	(Inch Units, Absolute programming)
N20 T0202	(Tool change Tool#2, offset #2)
N30 M08	**(Coolant start)**
N40 G97 S1000 M3	(RPM speed, speed 1000, CW rotation)
N50 G0 X1. Z0.1	(Rapid move to X0 Z0.1)
N60 G1 Z0 F0.012	(Linear interpolation G1 to Z0, feed rate 0.012)
N70 G3 X1.30Z-0.65 I0 K-0.65 F.012	(90° CCW Circular interpolation X1.30 Z-0.65 I0 K-0.65, feed rate F.015)

N80 G1 Z1.15	(Linear interpolation Z1.15, same feed rate 0.012)
N90 G3 X1.3 Z-0.65 I0.17 K.25 F.012	(Partial CCW Circular interpolation X1.30 Z-0.65 I0.17
	K0.25 feed rate F.012)
N100 G1 X1.4 F0.15	(Linear interpolation X1.4 feed rate 0.012)
N110 G0 X2.0	(Rapid move to X2.0)
N120 Z3.0	(Rapid move to Z3.0)
N130 M09	**(Coolant off)**
N140 T0200	(Tool#1 offset cancel #00)
N150 M5	(Stop spindle)
N160 M30	(Program end)
%	(End flag)

M30 Program End and Reset to the Beginning

Format structure: M30

M30 defines the end of the program execution, terminated the program and reset to the beginning (return the program cursor to the begging of the program). It can be used when multiple identical parts are produced one after another. M30 is the last block of the CNC programs. After M30 the CNC controlled switches off all operations – spindle rotation all axis feed fate and rapid movements, etc.

M30 Command Example:

N240 M30   End of the program return to the beginning

Let make an example program for the part shown in Figure 4.49.

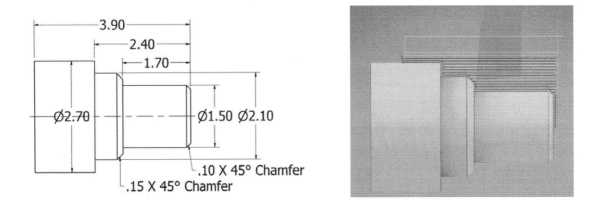

Figure 4.49 M30 example drawing and tool path

Example program for M30 (Example M30):

Tool: T0202 (tool #2, offset #2), Right hand general turning tool, insert shape C=80° degree
         Diamond, type T

Tool starting position: X3.5, Z3.5.

%	(Start flag)
o1130 (G00)	(Program Number 1130)
N10 G20 G90	(Inch Units, Absolute programming)
N20 T0202	(Tool change Tool#2, offset #2)
N30 M8	(Coolant start)
N40 G97 S650 M3	(RPM speed, speed 650, CW rotation)
N50 G0 X2.7 Z0.1	(Rapid move to X2.7 Z0.1)
N60 G1 Z-3.5 F0.013	(Linear interpolation G1 to Z-3.5, feed rate 0.013)
N70 G1 X2.8	(Linear interpolation G1 to X2.8, federate .013)
N80 G0 Z0.1	(Rapid move to Z0.1)
N90 X2.10	(Rapid move to X2.10)
N100 G1 Z-2.40	(Linear interpolation G1 to Z-2.40)
N110 G1 X2.80	(Linear interpolation G1 to X2.80)
N120 G00 Z0.1	(Rapid move to Z0.1)
N130 X1.5	(Rapid move to X1.5)
N140 G1 Z-1.70	(Linear interpolation G1 to Z-1.7)
N150 X 1.8	(Linear interpolation G1 to X1.8)
N160 X2.10 Z-1.85	(Linear interpolation G1 to X2.1 Z-1.85)
N170 X2.2	(Linear interpolation G1 to X2.2)
N180 G0 Z0	(Rapid move to Z0)
N190 G1 X0	(Linear interpolation G1 to X0)
N200 G1 Z0.1	(Linear interpolation G1 to Z0.1)
N210 G0 X1.3	(Rapid move to X1.3)
N220 G1 Z0	(Linear interpolation G1 to Z0)
N230 X1.5 Z-.10	(Linear interpolation G1 to X1.5 Z-0.1)
N240 X1.6	(Linear interpolation G1 to X1.6)
N250 G0 X 3.5	(Rapid move to X3.5)
N260 Z3.0	(Rapid move to Z3.0)
N270 M9	(Coolant off)
N280 T0100	(Tool#1 offset cancel #00)
N290M5	(Stop spindle)
N300 M30	**(Program end and reset to the beginning of the program)**
%	(End flag)

*Note: Feed rate (modal) specified during the first interpolation remain the same
    throughout the program unless changed.*

/ Block skip

Format structure: /

103

Skip the running of the same block after it is used. It must be specified on the most left position before the block number.

/ (Block Skip) Command Example:

/N120 G00 X1.3 Z-0.2   Block Skip, Rapid move to grove start position X1.3 Z-0.2

Block skip, also called block delete, is used to skip the running of the block after it is used. I is active when the block skip button on the CNC control panel is pressed. This option is used to program group of similar parts with a small differences in futures are machined. For example, block skip can be used for the section of the program than need to be omitted when certain future is no needed to be machined, and can be activated when is desired. If the block skip button is not activated, the program is executed in the regular sequence.

Let make an example program for the part shown in Figure 4.50.

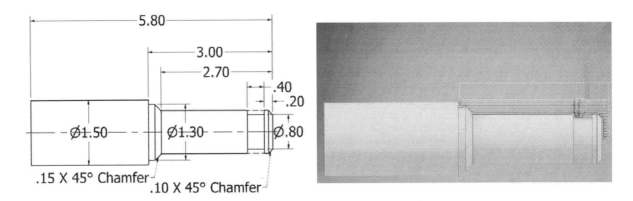

Figure 4.50 Block Skip example drawing and tool path

Example program for Block Skip (Example Block Skip):

Tool: T0101 (tool #1, offset #1), Right hand general turning tool, insert shape C=80° degree Diamond, type T

T1717 (tool #17, offset #17), Right hand grove tool, square profile, width 1/4" C=90° deg[13].

Tool starting position: X3.5, Z3.5.

%	(Start flag)
o1101	(Program Number 1101)
N10 G20 G90	(Inch Units, Absolute programming)
N20 T0101	(Tool change Tool#1, offset #1)
N30 M8	(Coolant start)
N40 G97 S1000 M3	(RPM speed, speed 500, CW rotation)
N50 G0 X1.0 Z0.1	(Rapid move to X3.0 Z0.1)

N60 G1 Z-2.7 F0.012	(Linear interpolation G1 to Z-2.7, feed rate 0.012)
N70 G1 X1.30 Z-2.85	(Linear interpolation G1 to X1.3, Z-2.85, federate .012)
N80 G28 X4.0 Z3.0	(G28 to home position passing through X4.0, Z3.0)
N90 T0100	(Tool#1 offset cancel #00)
/N100 T1717	(Block Skip, Tool change Tool#17, offset #17)
/N110 G29 X4.0 Z3.0	(Block Skip, G29 Return to start position passing through X4, Z3, Result: tool moved back to X1.3, Z-2.85)
/N120 G00 X1.3 Z-0.2	(Block Skip, Rapid move to grove start position X1.3 Z-0.2)
/N130 G75 X.80 Z-.475 F.005 I250 K0	(Block Skip, Grooving cycle X0.80, Z-4.75, feed rate F.005, I250 peck .025, K0 overstep)
/N140 G00 X3.5	(Block Skip, Rapid move to X3.5)
/N150 Z3.5	Block Skip, Rapid move to Z3.5)
/N160 T1700	(Tool#17 offset cancel #00)
N170 M09	(Coolant off)
N180 M5	(Stop spindle)
N190 M30	(Program end)
%	(End flag)

*Notes:

1. Skip block N100 to N160 active only when the skip button on CNC control panel is pressed, then the grove feature beneath the dashed line will be machine

2. Feed rate (modal) specified during the first interpolation remain the same throughout the program unless changed

---

(_ _ _ _) Comments

(_ _ _ _) Format structure: (Comment text)

Comments are placed between parenthesis "()" when there is a need to explain the program block or add additional information on running or setup. It can be specified on any block when needed. It is useful practice, similar to the one used by computer code programmers, which can make the program readable easy to understand and run. Properly used, the comment is completely ignored by the CNC controller, even if contains G and M code inside the text, and it is treated same as same as empty spaces inside the program.

(_ _ _ _) Comment Command Example:

N20 T0202   **(Tool change Tool#2, offset #2)**

Let make an example program for the part shown in Figure 4.51.

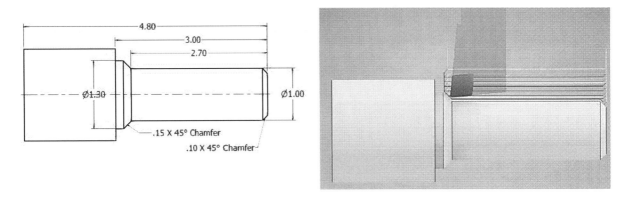

Figure 4.51 Comments example drawing and tool path

Example program for Comments (Example Comments):

Tool: T0202 (tool #2, offset #2), Right hand general turning tool, insert shape C=80° degree
Diamond, type T

Tool starting position: X3.5, Z3.5.

%	(Start flag)
o1102	(Program Number 1102)
N10 G20 G90	(Inch Units, Absolute programming)
N20 T0202	(Tool change Tool#2, offset #2)
N30 M8	(Coolant start)
N40 G97 S1000 M3	(RPM speed, speed 500, CW rotation)
N50 G0 X1.0 Z0.1	(Rapid move to X3.0 Z0.1)
N60 G1 Z-2.70 F0.012	(Linear interpolation G1 to Z-2.70, feed rate 0.012)
N70 G1 X1.30 Z-2.85	(Linear interpolation G1 to X1.3, Z-2.85)
N80 G1 Z-3.0	(Linear interpolation G1 to Z-3.0)
N90 G1 X 3.1	(Linear interpolation G1 to X3.1)
N100 G00 X3.5 Z3.5	(Rapid move to X3.5 Z3.5)
N110 T0200	(Tool#2 offset cancel #00)
N120 M09	(Coolant off)
N130 M5	(Stop spindle)
N140 M30	(Program end)
%	(End flag)

*Notes:

1. After each block comment with text placed between parenthesis ( ) can be added. Comments are ignored by the CNC controller and are treated as empty spaces

2. Feed rate (modal) specified during the first interpolation remain the same throughout the program unless changed

106

Notes:

Notes:

Chapter 5

CNC

Turning Examples

This chapter shows many turning drawing example, setup sheets, and programs. Some the samples have quite complex design and CNC programs for them are long. The programs are automatically created using Inventor HSM 2017 professional. It will benefit the user to work on each sample individually and try to create a simple program for several surfaces using advance G core cycles such as G71. G72, G70, G74, G75, G76 instead of multiple G00, G01, G02/G03. Then he can compare the results with automatically created programs as listed. 3D models (in formats like Inventor 2107, STEP, IGES) of the examples, drawings, and programs will be available for download via link upon request only for the users who purchased paper or electronic copy of the book.

Note: All dimensions and programs are in mm

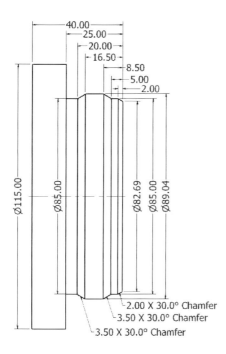 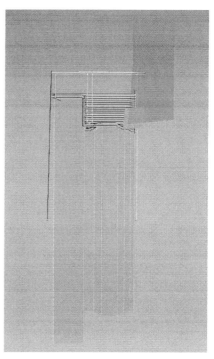

Figure 5.1 Drawing and tool path for part 9003

```
%
O9003 (Part 9003)
N10 G98 G18
N11 G21
N12 G50 S6000
N13 M31
N14 G53 G0 X0.

(Face5)
N15 T202
N16 G98
N17 M22
N18 G97 S500 M3
N19 G54
N20 M8
N21 G0 X140. Z5.
N22 G0 Z1.414
N23 G1 X122.828 F1000.
N24 G18
N25 G1 G41 X121.6 Z0.
N26 X-1.6
N28 G40 X1.228 Z1.414
N29 G0 X140.
N30 Z5.

(Profile3)
N31 G1 X138.4 Z5. F5000.
N32 G0 Z0.614
N33 X121.226
N34 G1 X119.228 F1000.
N36 G42 X116.4 Z-0.8
N37 Z-40.8
N39 G40 X119.228 Z-39.386
N40 G0 Z0.614
```

```
N41 X117.228
N43 G1 G42 X114.4 Z-0.8
F1000.
N44 Z-24.934
N45 G3 X115.4 Z-25.8 R1.
N46 G1 Z-40.8
N48 G40 X118.228 Z-39.386
N49 G0 Z0.614
N50 X115.228
N52 G1 G42 X112.4 Z-0.8
F1000.
N53 Z-24.8
N54 X113.4
N55 G3 X115.4 Z-25.798
R1.
N57 G1 G40 X118.228 Z-
24.384
N58 G0 Z0.614
N59 X113.228
N61 G1 G42 X110.4 Z-0.8
F1000.
N62 Z-24.8
N63 X113.4
N65 G40 X116.228 Z-23.386
N66 G0 Z0.614
N67 X111.228
N69 G1 G42 X108.4 Z-0.8
F1000.
N70 Z-24.8
N71 X111.4
N73 G40 X114.228 Z-23.386
N74 G0 Z0.614
N75 X109.228
```

```
N77 G1 G42 X106.4 Z-0.8
F1000.
N78 Z-24.8
N79 X109.4
N81 G40 X112.228 Z-23.386
N82 G0 Z0.614
N83 X107.228
N85 G1 G42 X104.4 Z-0.8
F1000.
N86 Z-24.8
N87 X107.4
N89 G40 X110.228 Z-23.386
N90 G0 Z0.614
N91 X105.228
N93 G1 G42 X102.4 Z-0.8
F1000.
N94 Z-24.8
N95 X105.4
N97 G40 X108.228 Z-23.386
N98 G0 Z0.614
N99 X103.228
N101 G1 G42 X100.4 Z-0.8
F1000.
N102 Z-24.8
N103 X103.4
N105 G40 X106.228 Z-
23.386
N106 G0 Z0.614
N107 X101.228
N109 G1 G42 X98.4 Z-0.8
F1000.
N110 Z-24.8
N111 X101.4
```

111

```
N113 G40 X104.228 Z-           N169 G0 Z0.614                 N225 G0 Z-21.333
23.386                         N170 X87.728                   N226 X92.602
N114 G0 Z0.614                 N172 G1 G42 X84.9 Z-0.8        N227 G1 X92.587 F1000.
N115 X99.228                   F1000.                         N228 X88.587
N117 G1 G42 X96.4 Z-0.8        N173 Z-2.099                   N229 X88.558
F1000.                         N174 X85.132 Z-2.3             N230 X92.558
N118 Z-24.8                    N175 G3 X85.4 Z-2.8 R1.        N231 G0 Z-22.25
N119 X99.4                     N176 G1 Z-5.532                N232 G1 X88.426 F1000.
N121 G40 X102.228 Z-           N177 X87.4 Z-7.264             N233 X87.499
23.386                         N179 G40 X90.228 Z-5.85        N234 X92.266
N122 G0 Z0.614                 N180 G0 Z0.614                 N235 G0 Z-23.167
N123 X97.228                   N181 X86.386                   N236 G1 X88.266 F1000.
N125 G1 G42 X94.4 Z-0.8        N182 G1 X86.228 F1000.         N237 X87.
F1000.                         N184 G42 X83.4 Z-0.8           N238 X91.42
N126 Z-24.8                    N185 X85.132 Z-2.3             N239 G0 Z-19.889
N127 X97.4                     N186 G3 X85.4 Z-2.8 R1.        N240 G1 X88.592 Z-21.304
N129 G40 X100.228 Z-           N188 G1 G40 X88.228 Z-         F1000.
23.386                         1.386                          N241 X87. Z-22.682
N130 G0 Z0.614                 N189 X88.69                    N242 Z-23.341
N131 X95.228                   N190 G0 X92.097                N243 X88.12
N133 G1 G42 X92.4 Z-0.8        N191 Z-16.918                  N244 Z-24.
F1000.                         N193 G1 G42 X89.268 Z-         N245 X87.
N134 Z-24.8                    18.333 F1000.                  N246 Z-23.341
N135 X95.4                     N194 X88.137 Z-24.8            N247 X88.365
N137 G40 X98.228 Z-23.386      N195 X90.268                   N248 Z-19.5
N138 G0 Z0.614                 N197 G40 X93.097 Z-23.386      N249 X85. Z-22.414
N139 X93.228                   N198 G0 Z0.614                 N250 Z-25.
N141 G1 G42 X90.4 Z-0.8        N199 X83.919                   N251 X87.945
F1000.                         N201 G1 G42 X81.091 Z-0.8      N252 X90.774 Z-23.586
N142 Z-24.8                    F1000.                         N253 G0 X140.
N143 X93.4                     N202 X83.4 Z-2.8               N254 Z5.
N145 G40 X96.228 Z-23.386      N203 Z-5.8                     N255 G53 X0.
N146 G0 Z0.614                 N204 X87.441 Z-9.3
N147 X91.228                   N205 Z-17.3                    (Part1)
N149 G1 G42 X88.4 Z-0.8        N206 X85.954 Z-25.8            N256 M9
F1000.                         N207 X113.4                    N257 M1
N150 Z-8.13                    N208 Z-40.8                    N258 T404
N151 X89.173 Z-8.8             N209 Z-35.785                  N259 G98
N152 G3 X89.441 Z-9.3 R1.      N210 Z-40.8                    N260 M22
N153 G1 Z-17.3                 N212 G40 X116.228 Z-           N261 G97 S500 M3
N154 G3 X89.434 Z-17.387       39.386                         N262 G54
R1.                            N213 G0 X138.4                 N263 M8
N155 G1 X89.268 Z-18.333       N214 Z5.                       N264 G0 X140. Z5.
N156 Z-24.8                    N215 G53 X0.                   N265 G0 Z-41.5
N157 X91.4                                                    N266 G18
N159 G40 X94.228 Z-23.386      (Groove2)                      N267 G1 G42 X-1.6 F1000.
N160 G0 Z0.614                 N216 M9                        N268 X140.
N161 X89.228                   N217 M1                        N269 G0 G40 Z5.
N163 G1 G42 X86.4 Z-0.8        N218 T303                      N270 Z5.
F1000.                         N219 G98
N164 Z-6.398                   N220 M22                       N271 M9
N165 X89.173 Z-8.8             N221 G97 S500 M3               N272 G53 X0.
N166 G3 X89.4 Z-9.098 R1.      N222 G54                       N273 G53 Z0.
N168 G1 G40 X92.228 Z-         N223 M8                        N274 M30
7.683                          N224 G0 X140. Z5.              %
```

Setup Sheet - Part 9003

Job

WCS: #0

Stock:

DX: 120mm
DY: 120mm
DZ: 80mm

Part:

DX: 115mm
DY: 115mm
DZ: 40mm

Stock Lower in WCS #0:
X: -60mm

Total

Number Of Operations: 4
Number Of Tools: 3
Tools: T2 T3 T4
Maximum Z: 5mm
Minimum Z: -41.5mm

Maximum Feedrate: 1000mm/min
Maximum Spindle Speed: 500rpm
Cutting Distance: 811.24mm
Rapid Distance: 838.31mm

Operation 1/4		
Description: Face5	Maximum Z: 5mm	T2 D0 L0
Strategy: Unspecified	Minimum Z: 0mm	Type: general turning
WCS: #0	Maximum Spindle Speed: 500rpm	Diameter: 0mm
Tolerance: 0.01mm	Maximum Feedrate: 1000mm/min	Length: 0mm
Compensation: control (center)	Cutting Distance: 73.15mm	Flutes: 1
Safe Tool Diameter: < 0mm	Rapid Distance: 76.56mm	
	Estimated Cycle Time: 5s (5.1%)	
	Coolant: Flood	

Operation 2/4		
Description: Profile3	Maximum Z: 5mm	T2 D0 L0
Strategy: Unspecified	Minimum Z: -40.8mm	Type: general turning
WCS: #0	Maximum Spindle Speed: 500rpm	Diameter: 0mm
Tolerance: 0.01mm	Maximum Feedrate: 1000mm/min	Length: 0mm
Stock to Leave: 0mm	Cutting Distance: 562.87mm	Flutes: 1
Maximum stepdown: 1mm	Rapid Distance: 561.2mm	
Maximum stepover: 1mm	Estimated Cycle Time: 41s (39%)	
Compensation: control (center)	Coolant: Flood	
Safe Tool Diameter: < 0mm		

Operation 3/4

DESCRIPTION: Groove2

STRATEGY: Unspecified

WCS: #0

MAXIMUM Z: 5mm

MINIMUM Z: -25mm

MAXIMUM SPINDLE SPEED: 500rpm
MAXIMUM FEEDRATE: 1000mm/min
CUTTING DISTANCE: 34.41mm
RAPID DISTANCE: 108.34mm

T3 D0 L0

TYPE: groove turning
DIAMETER: 0mm
LENGTH: 0mm

Operation 4/4

DESCRIPTION: Part1

STRATEGY: Unspecified

WCS: #0

TOLERANCE: 0.01mm

MAXIMUM Z: 5mm

MINIMUM Z: -41.5mm

MAXIMUM SPINDLE SPEED: 500rpm

MAXIMUM FEEDRATE: 1000mm/min

T4 D0 L0

TYPE: groove turning

DIAMETER: 0mm

LENGTH: 0mm

(Generated Inventor HSM Pro 4.0.0.032)

Part 0345

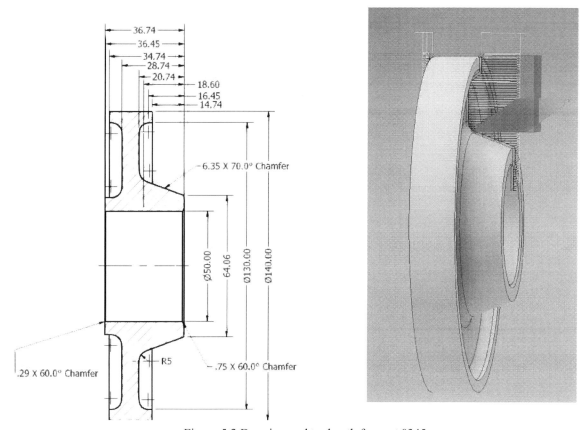

Figure 5.2 Drawing and tool path for part 0345

```
%
O0345 (Part 0345)
N10 G98 G18
N11 G21
N12 G50 S6000
N13 G28 U0.

(FACE1)
N14 T0101
N15 G54
N16 M8
N17 G98
N18 G97 S500 M3
N19 G0 X160. Z5.
N20 G0 Z-0.217
N21 G1 X142.828 F1000.
N22 X140. Z-1.631
N23 X-1.6
N24 X1.228 Z-0.217
N25 G0 X160.
N26 Z5.
N27 G28 U0.

(PROFILE1)
N28 M1
N29 T0202
N30 G54
N31 G98
N32 G97 S500 M3
```

```
N33 G0 X160. Z5.
N34 G0 Z1.404
N35 X141.821
N36 G1 X140.828 F1000.
N37 X138. Z-0.01
N38 Z-2.631
N39 X138.4
N40 G18 G3 X139.98 Z-2.814
R1.8
N41 G1 X142.809 Z-1.4
N42 G0 Z1.404
N43 X138.828
N44 G1 X136. Z-0.01 F1000.
N45 Z-2.631
N46 X138.4
N47 G3 X139. Z-2.656 R1.8
N48 G1 X141.828 Z-1.242
N49 G0 Z1.404
N50 X136.828
N51 G1 X134. Z-0.01 F1000.
N52 Z-2.631
N53 X137.
N54 X139.828 Z-1.217
N55 G0 Z1.404
N56 X134.828
N57 G1 X132. Z-0.01 F1000.
N58 Z-2.631
N59 X135.
N60 X137.828 Z-1.217
```

```
N61 G0 Z1.404
N62 X132.828
N63 G1 X130. Z-0.01 F1000.
N64 Z-2.631
N65 X133.
N66 X135.828 Z-1.217
N67 G0 Z1.404
N68 X130.828
N69 G1 X128. Z-0.01 F1000.
N70 Z-2.642
N71 G3 X128.4 Z-2.631 R1.8
N72 G1 X131.
N73 X133.828 Z-1.217
N74 G0 Z1.404
N75 X128.828
N76 G1 X126. Z-0.01 F1000.
N77 Z-3.09
N78 G3 X128.4 Z-2.631 R1.8
N79 G1 X129.
N80 X131.828 Z-1.217
N81 G0 Z1.404
N82 X126.828
N83 G1 X124. Z-0.01 F1000.
N84 Z-4.435
N85 X125.451 Z-3.399
N86 G3 X127. Z-2.773 R1.8
N87 G1 X129.828 Z-1.359
N88 G0 Z1.404
N89 X124.828
```

```
N90 G1 X122. Z-0.01 F1000.
N91 Z-5.863
N92 X125. Z-3.721
N93 G0 Z1.404
N94 X122.828
N95 G1 X120. Z-0.01 F1000.
N96 Z-7.291
N97 X123. Z-5.149
N98 G0 Z1.404
N99 X120.828
N100 G1 X118. Z-0.01 F1000.
N101 Z-8.631
N102 X118.123
N103 X121. Z-6.577
N104 G0 Z1.404
N105 X118.828
N106 G1 X116. Z-0.01 F1000.
N107 Z-8.631
N108 X118.123
N109 X119. Z-8.005
N110 G0 Z1.404
N111 X116.828
N112 G1 X114. Z-0.01 F1000.
N113 Z-8.631
N114 X117.
N115 X119.828 Z-7.217
N116 G0 Z1.404
N117 X114.828
N118 G1 X112. Z-0.01 F1000.
N119 Z-8.631
N120 X115.
N121 X117.828 Z-7.217
N122 G0 Z1.404
N123 X112.828
N124 G1 X110. Z-0.01 F1000.
N125 Z-8.631
N126 X113.
N127 X115.828 Z-7.217
N128 G0 Z1.404
N129 X110.828
N130 G1 X108. Z-0.01 F1000.
N131 Z-8.631
N132 X111.
N133 X113.828 Z-7.217
N134 G0 Z1.404
N135 X108.828
N136 G1 X106. Z-0.01 F1000.
N137 Z-8.631
N138 X109.
N139 X111.828 Z-7.217
N140 G0 Z1.404
N141 X106.828
N142 G1 X104. Z-0.01 F1000.
N143 Z-8.631
N144 X107.
N145 X109.828 Z-7.217
N146 G0 Z1.404
N147 X104.828
N148 G1 X102. Z-0.01 F1000.
N149 Z-8.631
N150 X105.
N151 X107.828 Z-7.217
N152 G0 Z1.404
N153 X102.828
N154 G1 X100. Z-0.01 F1000.
N155 Z-8.631
N156 X103.
N157 X105.828 Z-7.217
N158 G0 Z1.404
N159 X100.828
N160 G1 X98. Z-0.01 F1000.
N161 Z-8.631
N162 X101.
N163 X103.828 Z-7.217
N164 G0 Z1.404
N165 X98.828
N166 G1 X96. Z-0.01 F1000.
N167 Z-8.631
N168 X99.
N169 X101.828 Z-7.217
N170 G0 Z1.404
N171 X96.828
N172 G1 X94. Z-0.01 F1000.
N173 Z-8.631
N174 X97.
N175 X99.828 Z-7.217
N176 G0 Z1.404
N177 X94.828
N178 G1 X92. Z-0.01 F1000.
N179 Z-8.631
N180 X95.
N181 X97.828 Z-7.217
N182 G0 Z1.404
N183 X92.828
N184 G1 X90. Z-0.01 F1000.
N185 Z-8.631
N186 X93.
N187 X95.828 Z-7.217
N188 G0 Z1.404
N189 X90.828
N190 G1 X88. Z-0.01 F1000.
N191 Z-8.631
N192 X91.
N193 X93.828 Z-7.217
N194 G0 Z1.404
N195 X88.828
N196 G1 X86. Z-0.01 F1000.
N197 Z-8.631
N198 X89.
N199 X91.828 Z-7.217
N200 G0 Z1.404
N201 X86.828
N202 G1 X84. Z-0.01 F1000.
N203 Z-8.631
N204 X87.
N205 X89.828 Z-7.217
N206 G0 Z1.404
N207 X84.828
N208 G1 X82. Z-0.01 F1000.
N209 Z-8.631
N210 X85.
N211 X87.828 Z-7.217
N212 G0 Z1.404
N213 X82.828
N214 G1 X80. Z-0.01 F1000.
N215 Z-8.631
N216 X83.
N217 X85.828 Z-7.217
N218 G0 Z1.404
N219 X80.828
N220 G1 X78. Z-0.01 F1000.
N221 Z-8.631
N222 X81.
N223 X83.828 Z-7.217
N224 G0 Z1.404
N225 X78.828
N226 G1 X76. Z-0.01 F1000.
N227 Z-8.631
N228 X79.
N229 X81.828 Z-7.217
N230 G0 Z1.404
N231 X76.828
N232 G1 X74. Z-0.01 F1000.
N233 Z-8.631
N234 X77.
N235 X79.828 Z-7.217
N236 G0 Z1.404
N237 X74.828
N238 G1 X72. Z-0.01 F1000.
N239 Z-8.631
N240 X75.
N241 X77.828 Z-7.217
N242 G0 Z1.404
N243 X72.828
N244 G1 X70. Z-0.01 F1000.
N245 Z-8.625
N246 G2 X70.4 Z-8.631 R3.2
N247 G1 X73.
N248 X75.828 Z-7.217
N249 G0 Z1.404
N250 X70.828
N251 G1 X68. Z-0.01 F1000.
N252 Z-8.398
N253 G2 X70.4 Z-8.631 R3.2
N254 G1 X71.
N255 X73.828 Z-7.217
N256 G0 Z1.404
N257 X68.828
N258 G1 X66. Z-0.01 F1000.
N259 Z-7.755
N260 G2 X69. Z-8.554 R3.2
N261 G1 X71.828 Z-7.14
N262 G0 Z1.404
N263 X66.828
N264 G1 X64. Z-0.01 F1000.
N265 Z-2.42
N266 G3 Z-2.431 R1.8
N267 G1 Z-5.431
N268 G2 X67. Z-8.142 R3.2
N269 G1 X69.828 Z-6.728
N270 G0 Z1.404
N271 X64.828
N272 G1 X62. Z-0.01 F1000.
N273 Z-0.819
N274 G3 X64. Z-2.431 R1.8
N275 G1 X66.828 Z-1.017
N276 G0 Z1.404
N277 X62.828
N278 G1 X60. Z-0.01 F1000.
N279 Z-0.631
N280 X60.4
N281 G3 X63. Z-1.186 R1.8
N282 G1 X65.828 Z0.228
N283 G0 Z1.404
N284 X60.828
N285 G1 X58. Z-0.01 F1000.
N286 Z-0.631
N287 X60.4
N288 G3 X61. Z-0.656 R1.8
N289 G1 X63.828 Z0.758
N290 G0 Z1.404
N291 X58.828
N292 G1 X56. Z-0.01 F1000.
N293 Z-0.631
N294 X59.
N295 X61.828 Z0.783
N296 G0 Z1.404
N297 X56.828
N298 G1 X54. Z-0.01 F1000.
N299 Z-0.631
N300 X57.
N301 X59.828 Z0.783
N302 G0 Z1.404
N303 X54.828
N304 G1 X52. Z-0.01 F1000.
N305 Z-0.631
N306 X55.
N307 X57.828 Z0.783
N308 G0 Z1.404
```

```
N309 X52.828                  N382 X34.028 Z0.494           N450 Z-15.827
N310 G1 X50. Z-0.01 F1000.    N383 G0 Z-0.506               N451 G3 X128.4 Z-15.369 R1.8
N311 Z-0.631                  N384 G1 X31.228 F1000.        N452 G1 X129.
N312 X53.                     N385 X28.4 Z-1.92             N453 X131.828 Z-13.955
N313 X55.828 Z0.783           N386 X47.971                  N454 G0 Z1.404
N314 G0 Z1.404                N387 X48.6 Z-1.738            N455 X126.828
N315 X50.828                  N388 G3 X49.4 Z-1.631 R0.8    N456 G1 X124. Z-0.01 F1000.
N316 G1 X48. Z-0.01 F1000.    N389 X60.4                    N457 Z-17.173
N317 Z-0.773                  N390 G3 X62. Z-2.431 R0.8     N458 X125.451 Z-16.136
N318 G3 X49.4 Z-0.631 R1.8    N391 G1 Z-5.431               N459 G3 X127. Z-15.51 R1.8
N319 G1 X51.                  N392 G2 X70.4 Z-9.631 R4.2    N460 G1 X129.828 Z-14.096
N320 X53.828 Z0.783           N393 G1 X118.4                N461 G0 Z1.404
N321 G0 Z1.404                N394 X118.401                 N462 X124.828
N322 X48.828                  N395  G2  X119.191  Z-9.613   N463 G1 X122. Z-0.01 F1000.
N323 G1 X46. Z-0.01 F1000.    R4.2                          N464 Z-18.601
N324 Z-0.92                   N396 G1 X127.089 Z-3.972      N465 X125. Z-16.458
N325 X47.435                  N397 G3 X128.4 Z-3.631 R0.8   N466 G0 Z1.404
N326 X47.6 Z-0.872            N398 G1 X138.4                N467 X122.828
N327 G3 X49. Z-0.642 R1.8     N399  G3  X139.982  Z-4.311   N468 G1 X120. Z-0.01 F1000.
N328 G1 X51.828 Z0.772        R0.8                          N469 Z-20.029
N329 G0 Z1.404                N400 G1 X142.81 Z-2.897       N470 X123. Z-17.887
N330 X46.828                  N401 X143.084                 N471 G0 Z1.404
N331 G1 X44. Z-0.01 F1000.    N402 G0 X160.                 N472 X120.828
N332 Z-0.92                   N403 Z5.                      N473 G1 X118. Z-0.01 F1000.
N333 X47.                                                   N474 Z-21.369
N334 X49.828 Z0.494           (PROFILE2)                    N475 X118.123
N335 G0 Z1.404                N404 G98                      N476 X121. Z-19.315
N336 X44.828                  N405 G97 S500 M3              N477 G0 Z1.404
N337 G1 X42. Z-0.01 F1000.    N406 G0 X160. Z5.             N478 X118.828
N338 Z-0.92                   N407 Z1.404                   N479 G1 X116. Z-0.01 F1000.
N339 X45.                     N408 X141.821                 N480 Z-21.369
N340 X47.828 Z0.494           N409 G1 X140.828 F1000.       N481 X118.123
N341 G0 Z1.404                N410 X138. Z-0.01             N482 X119. Z-20.743
N342 X42.828                  N411 Z-15.369                 N483 G0 Z1.404
N343 G1 X40. Z-0.01 F1000.    N412 X138.4                   N484 X116.828
N344 Z-0.92                   N413  G3  X139.98  Z-15.551   N485 G1 X114. Z-0.01 F1000.
N345 X43.                     R1.8                          N486 Z-21.369
N346 X45.828 Z0.494           N414 G1 X142.809 Z-14.137     N487 X117.
N347 G0 Z1.404                N415 G0 Z1.404                N488 X119.828 Z-19.955
N348 X40.828                  N416 X138.828                 N489 G0 Z1.404
N349 G1 X38. Z-0.01 F1000.    N417 G1 X136. Z-0.01 F1000.   N490 X114.828
N350 Z-0.92                   N418 Z-15.369                 N491 G1 X112. Z-0.01 F1000.
N351 X41.                     N419 X138.4                   N492 Z-21.369
N352 X43.828 Z0.494           N420 G3 X139. Z-15.394 R1.8   N493 X115.
N353 G0 Z1.404                N421 G1 X141.828 Z-13.98      N494 X117.828 Z-19.955
N354 X38.828                  N422 G0 Z1.404                N495 G0 Z1.404
N355 G1 X36. Z-0.01 F1000.    N423 X136.828                 N496 X112.828
N356 Z-0.92                   N424 G1 X134. Z-0.01 F1000.   N497 G1 X110. Z-0.01 F1000.
N357 X39.                     N425 Z-15.369                 N498 Z-21.369
N358 X41.828 Z0.494           N426 X137.                    N499 X113.
N359 G0 Z1.404                N427 X139.828 Z-13.955        N500 X115.828 Z-19.955
N360 X36.828                  N428 G0 Z1.404                N501 G0 Z1.404
N361 G1 X34. Z-0.01 F1000.    N429 X134.828                 N502 X110.828
N362 Z-0.92                   N430 G1 X132. Z-0.01 F1000.   N503 G1 X108. Z-0.01 F1000.
N363 X37.                     N431 Z-15.369                 N504 Z-21.369
N364 X39.828 Z0.494           N432 X135.                    N505 X111.
N365 G0 Z1.404                N433 X137.828 Z-13.955        N506 X113.828 Z-19.955
N366 X34.828                  N434 G0 Z1.404                N507 G0 Z1.404
N367 G1 X32. Z-0.01 F1000.    N435 X132.828                 N508 X108.828
N368 Z-0.92                   N436 G1 X130. Z-0.01 F1000.   N509 G1 X106. Z-0.01 F1000.
N369 X35.                     N437 Z-15.369                 N510 Z-21.369
N370 X37.828 Z0.494           N438 X133.                    N511 X109.
N371 G0 Z1.404                N439 X135.828 Z-13.955        N512 X111.828 Z-19.955
N372 X33.028                  N440 G0 Z1.404                N513 G0 Z1.404
N373 G1 X30.2 Z-0.01 F1000.   N441 X130.828                 N514 X106.828
N374 Z-0.92                   N442 G1 X128. Z-0.01 F1000.   N515 G1 X104. Z-0.01 F1000.
N375 X33.                     N443 Z-15.38                  N516 Z-21.369
N376 X35.828 Z0.494           N444 G3 X128.4 Z-15.369 R1.8  N517 X107.
N377 G0 Z1.404                N445 G1 X131.                 N518 X109.828 Z-19.955
N378 X31.228                  N446 X133.828 Z-13.955        N519 G0 Z1.404
N379 G1 X28.4 Z-0.01 F1000.   N447 G0 Z1.404                N520 X104.828
N380 Z-0.92                   N448 X128.828                 N521 G1 X102. Z-0.01 F1000.
N381 X31.2                    N449 G1 X126. Z-0.01 F1000.   N522 Z-21.369
```

```
N523 X105.
N524 X107.828 Z-19.955
N525 G0 Z1.404
N526 X102.828
N527 G1 X100. Z-0.01 F1000.
N528 Z-21.369
N529 X103.
N530 X105.828 Z-19.955
N531 G0 Z1.404
N532 X100.828
N533 G1 X98. Z-0.01 F1000.
N534 Z-21.369
N535 X101.
N536 X103.828 Z-19.955
N537 G0 Z1.404
N538 X98.828
N539 G1 X96. Z-0.01 F1000.
N540 Z-21.369
N541 X99.
N542 X101.828 Z-19.955
N543 G0 Z1.404
N544 X96.828
N545 G1 X94. Z-0.01 F1000.
N546 Z-21.369
N547 X97.
N548 X99.828 Z-19.955
N549 G0 Z1.404
N550 X94.828
N551 G1 X92. Z-0.01 F1000.
N552 Z-21.369
N553 X95.
N554 X97.828 Z-19.955
N555 G0 Z1.404
N556 X92.828
N557 G1 X90. Z-0.01 F1000.
N558 Z-21.369
N559 X93.
N560 X95.828 Z-19.955
N561 G0 Z1.404
N562 X90.828
N563 G1 X88. Z-0.01 F1000.
N564 Z-21.369
N565 X91.
N566 X93.828 Z-19.955
N567 G0 Z1.404
N568 X88.828
N569 G1 X86. Z-0.01 F1000.
N570 Z-21.369
N571 X89.
N572 X91.828 Z-19.955
N573 G0 Z1.404
N574 X86.828
N575 G1 X84. Z-0.01 F1000.
N576 Z-21.356
N577   G2  X84.564  Z-21.369
R3.2
N578 G1 X87.
N579 X89.828 Z-19.955
N580 G0 Z1.404
N581 X84.828
N582 G1 X82. Z-0.01 F1000.
N583 Z-21.101
N584   G2  X84.564  Z-21.369
R3.2
N585 G1 X85.
N586 X87.828 Z-19.955
N587 G0 Z1.404
N588 X82.828
N589 G1 X80. Z-0.01 F1000.
N590 Z-20.412
N591 G2 X83. Z-21.272 R3.2
N592 G1 X85.828 Z-19.858
N593 G0 Z1.404

N594 X80.828
N595 G1 X78. Z-0.01 F1000.
N596 Z-18.508
N597 X78.55 Z-19.263
N598 G2 X81. Z-20.827 R3.2
N599 G1 X83.828 Z-19.413
N600 G0 Z1.404
N601 X78.828
N602 G1 X76. Z-0.01 F1000.
N603 Z-15.76
N604 X78.55 Z-19.263
N605 G2 X79. Z-19.75 R3.2
N606 G1 X81.828 Z-18.336
N607 G0 Z1.404
N608 X76.828
N609 G1 X74. Z-0.01 F1000.
N610 Z-13.013
N611 X77. Z-17.134
N612 X79.828 Z-15.72
N613 G0 Z1.404
N614 X74.828
N615 G1 X72. Z-0.01 F1000.
N616 Z-10.265
N617 X75. Z-14.387
N618 X77.828 Z-12.972
N619 G0 Z1.404
N620 X72.828
N621 G1 X70. Z-0.01 F1000.
N622 Z-7.518
N623 X73. Z-11.639
N624 X75.828 Z-10.225
N625 G0 Z1.404
N626 X70.828
N627 G1 X68. Z-0.01 F1000.
N628 Z-4.77
N629 X71. Z-8.892
N630 X73.828 Z-7.477
N631 G0 Z1.404
N632 X68.828
N633 G1 X66. Z-0.01 F1000.
N634 Z-2.023
N635 X69. Z-6.144
N636 X71.828 Z-4.73
N637 G0 Z1.404
N638 X66.828
N639 G1 X64. Z-0.01 F1000.
N640 Z-0.803
N641 G3 X65.849 Z-1.816 R1.8
N642 G1 X67. Z-3.397
N643 X69.828 Z-1.982
N644 G0 Z1.404
N645 X64.828
N646 G1 X62. Z-0.01 F1000.
N647 Z-0.631
N648 X62.466
N649 G3 X65. Z-1.153 R1.8
N650 G1 X67.828 Z0.262
N651 G0 Z1.404
N652 X62.828
N653 G1 X60. Z-0.01 F1000.
N654 Z-0.631
N655 X62.466
N656 G3 X63. Z-0.651 R1.8
N657 G1 X65.828 Z0.763
N658 G0 Z1.404
N659 X60.828
N660 G1 X58. Z-0.01 F1000.
N661 Z-0.631
N662 X61.
N663 X63.828 Z0.783
N664 G0 Z1.404
N665 X58.828
N666 G1 X56. Z-0.01 F1000.

N667 Z-0.631
N668 X59.
N669 X61.828 Z0.783
N670 G0 Z1.404
N671 X56.828
N672 G1 X54. Z-0.01 F1000.
N673 Z-0.631
N674 X57.
N675 X59.828 Z0.783
N676 G0 Z1.404
N677 X54.828
N678 G1 X52. Z-0.01 F1000.
N679 Z-0.631
N680 X55.
N681 X57.828 Z0.783
N682 G0 Z1.404
N683 X52.828
N684 G1 X50. Z-0.01 F1000.
N685 Z-0.702
N686 G3 X50.998 Z-0.631 R1.8
N687 G1 X53.
N688 X55.828 Z0.783
N689 G0 Z1.404
N690 X50.828
N691 G1 X48. Z-0.01 F1000.
N692 Z-1.218
N693 X49.198 Z-0.872
N694 G3 X50.998 Z-0.631 R1.8
N695 G1 X53.826 Z0.783
N696 G0 Z1.404
N697 X48.828
N698 G1 X46. Z-0.01 F1000.
N699 Z-1.381
N700 X47.435
N701 X49. Z-0.93
N702 X51.828 Z0.485
N703 G0 Z1.404
N704 X46.828
N705 G1 X44. Z-0.01 F1000.
N706 Z-1.381
N707 X47.
N708 X49.828 Z0.033
N709 G0 Z1.404
N710 X44.828
N711 G1 X42. Z-0.01 F1000.
N712 Z-1.381
N713 X45.
N714 X47.828 Z0.033
N715 G0 Z1.404
N716 X42.828
N717 G1 X40. Z-0.01 F1000.
N718 Z-1.381
N719 X43.
N720 X45.828 Z0.033
N721 G0 Z1.404
N722 X40.828
N723 G1 X38. Z-0.01 F1000.
N724 Z-1.381
N725 X41.
N726 X43.828 Z0.033
N727 G0 Z1.404
N728 X38.828
N729 G1 X36. Z-0.01 F1000.
N730 Z-1.381
N731 X39.
N732 X41.828 Z0.033
N733 G0 Z1.404
N734 X36.828
N735 G1 X34. Z-0.01 F1000.
N736 Z-1.381
N737 X37.
N738 X39.828 Z0.033
N739 G0 Z1.404
```

```
N740 X34.828                N755 X31.2                    N769 G3 X128.4 Z-16.369 R0.8
N741 G1 X32. Z-0.01 F1000.  N756 X34.028 Z0.033           N770 G1 X138.4
N742 Z-1.381                N757 G0 Z-0.967               N771  G3  X139.982  Z-17.049
N743 X35.                   N758 G1 X31.228 F1000.        R0.8
N744 X37.828 Z0.033         N759 X28.4 Z-2.381           N772 G1 X142.81 Z-15.635
N745 G0 Z1.404              N760 X47.971                  N773 X143.084
N746 X33.028                N761 X50.198 Z-1.738          N774 G0 X160.
N747 G1 X30.2 Z-0.01 F1000. N762 G3 X50.998 Z-1.631 R0.8  N775 Z5.
N748 Z-1.381                N763 G1 X62.466
N749 X33.                   N764 G3 X63.97 Z-2.158 R0.8    N776 M9
N750 X35.828 Z0.033         N765 G1 X76.671 Z-19.605      N777 G28 U0. W0.
N751 G0 Z1.404              N766  G2  X84.564  Z-22.369   N778 M30
N752 X31.228                R4.2                          %
N753 G1 X28.4 Z-0.01 F1000. N767 G1 X119.165
N754 Z-1.381                N768 X127.089 Z-16.71
```

Setup Sheet - Part 0345

Job

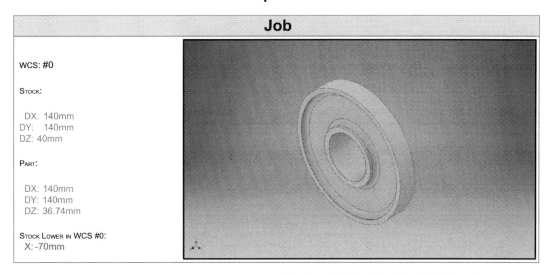

WCS: #0

STOCK:

DX: 140mm
DY: 140mm
DZ: 40mm

PART:

DX: 140mm
DY: 140mm
DZ: 36.74mm

STOCK LOWER IN WCS #0:
X: -70mm

Total

NUMBER OF OPERATIONS: 3 NUMBER OF TOOLS: 2
TOOLS: T1 T2

MAXIMUM Z: 5mm

MAXIMUM FEEDRATE: 1000mm/min MAXIMUM SPINDLE
SPEED: 500rpm CUTTING DISTANCE: 1834.21mm RAPID
DISTANCE: 1420.53mm ESTIMATED CYCLE TIME: 2m:37s

cration 1/3
SCRIPTION: Face1   ‹IMUM Z: 5mm   D0 L0
ATEGY: Unspecified   IMUM Z: -1.63mm   E: general turning
S: #0   ‹IMUM SPINDLE SPEED: 500rpm   METER: 0mm
ERANCE: 0.01mm   ‹IMUM FEEDRATE: 1000mm/min   GTH: 0mm
                           TING DISTANCE: 83.39mm   TES: 1
                           RAPID DISTANCE: 89.82mm

ESTIMATED CYCLE TIME: 6s (3.9%)

COOLANT: Flood

Operation 2/3

DESCRIPTION: Profile1	MAXIMUM Z: 5mm	**T2** D0 L0
STRATEGY: Unspecified	MINIMUM Z: -9.63mm	TYPE: boring turning
WCS: #0	MAXIMUM SPINDLE SPEED: 500rpm	DIAMETER: 0mm
TOLERANCE: 0.01mm	MAXIMUM FEEDRATE: 1000mm/min	LENGTH: 0mm
STOCK TO LEAVE: 0mm	CUTTING DISTANCE: 665.54mm	FLUTES: 1
MAXIMUM STEPDOWN: 1mm	RAPID DISTANCE: 449.81mm	
MAXIMUM STEPOVER: 1mm	ESTIMATED CYCLE TIME: 45s (28.9%)	
	COOLANT: Flood	

Operation 3/3

DESCRIPTION: Profile2	MAXIMUM Z: 5mm	**T2** D0 L0
STRATEGY: Unspecified	MINIMUM Z: -22.37mm	TYPE: boring turning
WCS: #0		DIAMETER: 0mm
	MAXIMUM SPINDLE SPEED: 500rpm	LENGTH: 0mm
TOLERANCE: 0.01mm	MAXIMUM FEEDRATE: 1000mm/min	
STOCK TO LEAVE: 0mm	CUTTING DISTANCE: 1085.29mm	
	RAPID DISTANCE: 880.9mm	

Generated by Inventor HSM Pro 4.0.0.032

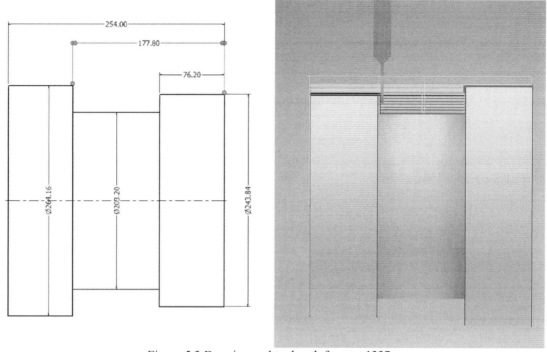

Figure 5.3 Drawing and tool path for part 1337

```
%
O1337 (Part 1337)
N10 G98 G18
N11 G21
N12 G50 S6000
N13 M31
N14 G53 G0 X0.

(Face4)
N15 T202
N16 G98
N17 M22
N18 G97 S500 M3
N19 G54
N20 M8
N21 G0 X287.02 Z5.
N22 G0 Z1.437
N23 G1 X269.574 F1000.
N24 X266.7 Z0.
N25 X-0.4
N26 X2.474 Z1.437
N27 G0 X287.02
N28 Z5.

(Profile3)
N29 M9
N30 G1 X287.02 Z5. F5000.
N31 G0 Z1.427
N32 X269.327
N33 G1 X267.542 F1000.
N34 X264.668 Z-0.01
N35 Z-176.872
N36 G18 G3 X266.192 Z-178.
R1.216
```

```
N37 G1 Z-254.2
N38 X266.68
N39 X269.554 Z-252.763
N40 G0 Z1.427
N41 X265.51
N42 G1 X262.636 Z-0.01
F1000.
N43 Z-176.784
N44 X263.76
N45 G3 X265.684 Z-177.256
R1.216
N46 G1 X268.558 Z-175.819
N47 G0 Z1.427
N48 X263.478
N49 G1 X260.604 Z-0.01
F1000.
N50 Z-176.784
N51 X263.652
N52 X266.526 Z-175.347
N53 G0 Z1.427
N54 X261.446
N55 G1 X258.572 Z-0.01
F1000.
N56 Z-176.784
N57 X261.62
N58 X264.494 Z-175.347
N59 G0 Z1.427
N60 X259.414
N61 G1 X256.54 Z-0.01
F1000.
N62 Z-176.784
N63 X259.588
N64 X262.462 Z-175.347
N65 G0 Z1.427
```

```
N66 X257.382
N67 G1 X254.508 Z-0.01
F1000.
N68 Z-176.784
N69 X257.556
N70 X260.43 Z-175.347
N71 G0 Z1.427
N72 X255.35
N73 G1 X252.476 Z-0.01
F1000.
N74 Z-176.784
N75 X255.524
N76 X258.398 Z-175.347
N77 G0 Z1.427
N78 X253.318
N79 G1 X250.444 Z-0.01
F1000.
N80 Z-176.784
N81 X253.492
N82 X256.366 Z-175.347
N83 G0 Z1.427
N84 X251.286
N85 G1 X248.412 Z-0.01
F1000.
N86 Z-176.784
N87 X251.46
N88 X254.334 Z-175.347
N89 G0 Z1.427
N90 X250.016
N91 G1 X247.142 Z-0.01
F1000.
N92 Z-176.784
N93 X249.428
N94 X252.302 Z-175.347
```

```
N95 G0 Z1.427                  N125 G1 X238.284               N160 X206.534
N96 X248.746                   N126 Z-177.8                   N161 Z-177.8
N97 G1 X245.872 Z-0.01         N127 X244.634                  N162 X212.884
F1000.                         N128 X247.507 Z-176.363        N163 X215.757 Z-176.363
N98 Z-176.784                  N129 G0 Z-79.248               N164 G0 Z-79.248
N99 X248.158                   N130 G1 X242.348 F1016.        N165 G1 X210.598 F1016.
N100 X251.032 Z-175.347        N131 X238.284                  N166 X206.534
N101 G0 Z1.237                 N132 X231.934                  N167 X203.994
N102 X246.714                  N133 Z-177.8                   N168 Z-177.8
N103 G1 X243.84 Z-0.2          N134 X238.284                  N169 X206.534
F1000.                         N135 X241.157 Z-176.363        N170 X209.407 Z-176.363
N104 Z-177.8                   N136 G0 Z-79.248               N171 X248.698
N105 X263.76                   N137 G1 X235.998 F1016.        N172 G0 Z-79.248
N106 G3 X264.16 Z-178. R0.2    N138 X231.934                  N173 G1 X244.634 F1016.
N107 G1 Z-254.2                N139 X225.584                  N174 X203.994
N108 X267.034 Z-252.763        N140 Z-177.8                   N175 Z-128.524
N109 X268.224                  N141 X231.934                  N176 X206.867 Z-127.087
N110 G0 X287.02                N142 X234.807 Z-176.363        N177 X208.058
N111 Z5.                       N143 G0 Z-79.248               N178 G0 X248.698
N112 G53 X0.                   N144 G1 X229.648 F1016.        N179 Z-177.8
                               N145 X225.584                  N180 G1 X244.634 F1016.
                               N146 X219.234                  N181 X203.994
(Groove4)                      N147 Z-177.8                   N182 Z-128.524
N113 M1                        N148 X225.584                  N183 X206.867 Z-129.961
N114 T506                      N149 X228.457 Z-176.363        N184 X208.058
N115 G98                       N150 G0 Z-79.248               N185 G0 X287.814
N116 M22                       N151 G1 X223.298 F1016.        N186 Z5.
N117 G97 S500 M3               N152 X219.234
N118 G54                       N153 X212.884                  N187 G53 X0.
N119 G0 X287.814 Z5.           N154 Z-177.8                   N188 G53 Z0.
N120 G0 Z-77.417               N155 X219.234                  N189 M30
N121 X248.698                  N156 X222.107 Z-176.363        %
N122 G1 X247.507 F1016.        N157 G0 Z-79.248
N123 X244.634 Z-78.854         N158 G1 X216.948 F1016.
N124 G18 G3 X243.84 Z-         N159 X212.884
79.248 R0.397
```

Setup Sheet - Part 1337

Job

WCS: #0

STOCK:

DX: 10.5in
DY: 10.5in
DZ: 10in

PART:

DX: 10.4in
DY: 10.4in
DZ: 10in

STOCK LOWER IN WCS #0:
X: -5.25in

Total

NUMBER OF OPERATIONS: 3
NUMBER OF TOOLS: 2
TOOLS: T2 T5

MAXIMUM Z: 0.197in

MINIMUM Z: -10.008in

MAXIMUM FEEDRATE: 40in/min
MAXIMUM SPINDLE SPEED: 500rpm

Operation 1/3

DESCRIPTION: Face4	MAXIMUM Z: 0.197in	**T2** D0 L0
STRATEGY: Unspecified	MINIMUM Z: 0in	TYPE: general turning
WCS: #0	MAXIMUM SPINDLE SPEED: 500rpm	DIAMETER: 0in
TOLERANCE: 0in	MAXIMUM FEEDRATE: 39.37in/min	LENGTH: 0in
	CUTTING DISTANCE: 5.761in	FLUTES: 1
	RAPID DISTANCE: 5.882in	
	ESTIMATED CYCLE TIME: 9s (3.7%)	
	COOLANT: Flood	

Operation 2/3

DESCRIPTION: Profile3	MAXIMUM Z: 0.197in	**T2** D0 L0
STRATEGY: Unspecified	MINIMUM Z: -10.008in	TYPE: general turning
WCS: #0	MAXIMUM SPINDLE SPEED: 500rpm	DIAMETER: 0in
TOLERANCE: 0in	MAXIMUM FEEDRATE: 39.37in/min	LENGTH: 0in
STOCK TO LEAVE: 0in	CUTTING DISTANCE: 92.577in	FLUTES: 1
MAXIMUM STEPDOWN: 0.04in	RAPID DISTANCE: 91.641in	
MAXIMUM STEPOVER: 0.04in	ESTIMATED CYCLE TIME: 2m:22s (59.9%)	
	COOLANT: Off	

Operation 3/3

DESCRIPTION: Groove4

STRATEGY: Unspecified

MAXIMUM Z: 0.197in

MINIMUM Z: -7in

MAXIMUM SPINDLE SPEED: 500rpm
MAXIMUM FEEDRATE: 40in/min
CUTTING DISTANCE: 37.142in
RAPID DISTANCE: 40.459in

T5 D0 L0

TYPE: groove turning

DIAMETER: 0in

Generated by Inventor HSM Pro 4.0.0.032

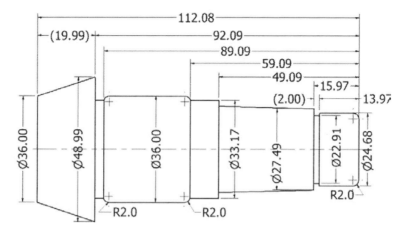

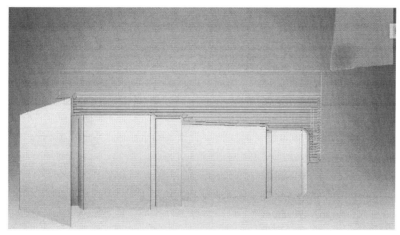

Figure 5.4 Drawing and tool path for part 9002

```
%
O9002 (Part 9002)
N10 G98 G18
N11 G21
N12 G50 S6000
N13 M31
N14 G53 G0 X0.

(Profile1)
N15 T101
N16 G98
N17 M22
N18 G97 S500 M3
N19 G54
N20 M8
N21 G0 X70. Z5.
N22 G0 Z1.404
N23 X51.821
N24 G1 X50.828 F1000.
N25 X48. Z-0.01
N26 Z-95.078
N27 G18 G3 X49.981 Z-95.602
R1.8
N28 G1 X52.809 Z-94.188
N29 G0 Z1.404
```

```
N30 X48.828
N31 G1 X46. Z-0.01 F1000.
N32 Z-95.052
N33 X47.389
N34 G3 X49. Z-95.242 R1.8
N35 G1 X51.828 Z-93.828
N36 G0 Z1.404
N37 X46.828
N38 G1 X44. Z-0.01 F1000.
N39 Z-95.052
N40 X47.
N41 X49.828 Z-93.637
N42 G0 Z1.404
N43 X44.828
N44 G1 X42. Z-0.01 F1000.
N45 Z-95.052
N46 X45.
N47 X47.828 Z-93.637
N48 G0 Z1.404
N49 X42.828
N50 G1 X40. Z-0.01 F1000.
N51 Z-95.052
N52 X43.
N53 X45.828 Z-93.637
N54 G0 Z1.404
```

```
N55 X40.828
N56 G1 X38. Z-0.01 F1000.
N57 Z-65.757
N58 Z-65.764
N59 Z-91.94
N60 G3 X37.971 Z-92.271 R3.8
N61 G1 X37.871 Z-92.842
N62 Z-95.052
N63 X41.
N64 X43.828 Z-93.637
N65 G0 Z1.404
N66 X38.828
N67 G1 X36. Z-0.01 F1000.
N68 Z-63.195
N69 G3 X38. Z-65.764 R3.8
N70 G1 X40.828 Z-64.35
N71 G0 Z1.404
N72 X36.828
N73 G1 X34. Z-0.01 F1000.
N74 Z-52.522
N75 G3 X35.173 Z-53.852 R1.8
N76 G1 Z-62.807
N77 G3 X37. Z-63.88 R3.8
N78 G1 X39.828 Z-62.465
N79 G0 Z1.404
```

```
N80  X34.828                      N131 Z-2.96                       N183 G0 Z1.404
N81  G1 X32. Z-0.01 F1000.        N132 X19.08                       N184 X2.828
N82  Z-52.064                     N133 G3 X21. Z-3.083 R3.8         N185 G1 X0. Z-0.01 F1000.
N83  G3 X35. Z-53.3 R1.8          N134 G1 X23.828 Z-1.669           N186 Z-2.96
N84  G1 X37.828 Z-51.886          N135 G0 Z1.404                    N187 X3.
N85  G0 Z1.404                    N136 X18.828                      N188 X5.828 Z-1.546
N86  X32.828                      N137 G1 X16. Z-0.01 F1000.        N189 X24.156
N87  G1 X30. Z-0.01 F1000.        N138 Z-2.96                       N190 G0 X40.7
N88  Z-30.713                     N139 X19.                         N191 Z-91.428
N89  X31.083 Z-52.052             N140 X21.828 Z-1.546              N192 G1 X37.871  Z-92.842
N90  X31.573                      N141 G0 Z1.404                    F1000.
N91  G3 X33. Z-52.199 R1.8        N142 X16.828                      N193 X37.484 Z-95.052
N92  G1 X35.828 Z-50.785          N143 G1 X14. Z-0.01 F1000.        N194 X38.871
N93  G0 Z1.404                    N144 Z-2.96                       N195 X41.7 Z-93.637
N94  X30.828                      N145 X17.                         N196 G0 Z-2.546
N95  G1 X28. Z-0.01 F1000.        N146 X19.828 Z-1.546              N197 X23.672
N96  Z-19.27                      N147 G0 Z1.404                    N198 G1 X1.228 F1000.
N97  G3 X29.491 Z-20.684 R1.8     N148 X14.828                      N199 X-1.6 Z-3.96
N98  G1 X31. Z-50.419             N149 G1 X12. Z-0.01 F1000.        N200 X19.08
N99  X33.828 Z-49.005             N150 Z-2.96                       N201 G3 X24.68 Z-6.76 R2.8
N100 G0 Z1.404                    N151 X15.                         N202 G1 Z-18.729
N101 X28.828                      N152 X17.828 Z-1.546              N203 G3  X24.674  Z-18.799
N102 G1 X26. Z-0.01 F1000.        N153 G0 Z1.404                    R0.8
N103 Z-5.189                      N154 X12.828                      N204 G1 X24.476 Z-19.929
N104 G3 X26.68 Z-6.76 R3.8        N155 G1 X10. Z-0.01 F1000.        N205 X25.892
N105 G1 Z-18.729                  N156 Z-2.96                       N206 G3  X27.492  Z-20.709
N106  G3  X26.666  Z-18.886       N157 X13.                         R0.8
R1.8                              N158 X15.828 Z-1.546              N207 G1 X29.133 Z-53.052
N107 G1 X26.651 Z-18.97           N159 G0 Z1.404                    N208 X31.573
N108 G3 X29. Z-19.821 R1.8        N160 X10.828                      N209  G3  X33.173  Z-53.852
N109 G1 X31.828 Z-18.407          N161 G1 X8. Z-0.01 F1000.         R0.8
N110 G0 Z1.404                    N162 Z-2.96                       N210 G1 Z-63.331
N111 X26.828                      N163 X11.                         N211 G3 X36. Z-65.764 R2.8
N112 G1 X24. Z-0.01 F1000.        N164 X13.828 Z-1.546              N212 G1 Z-91.94
N113 Z-3.864                      N165 G0 Z1.404                    N213  G3  X35.979  Z-92.184
N114 G3 X26.68 Z-6.76 R3.8        N166 X8.828                       R2.8
N115 G1 X29.508 Z-5.346           N167 G1 X6. Z-0.01 F1000.         N214 G1 X35.302 Z-96.052
N116 G0 Z1.404                    N168 Z-2.96                       N215 X47.389
N117 X24.828                      N169 X9.                          N216  G3  X48.983  Z-96.921
N118 G1 X22. Z-0.01 F1000.        N170 X11.828 Z-1.546              R0.8
N119 Z-3.252                      N171 G0 Z1.404                    N217 G1 X48.122 Z-101.84
N120 G3 X25. Z-4.377 R3.8         N172 X6.828                       N218 X50.951 Z-100.426
N121 G1 X27.828 Z-2.963           N173 G1 X4. Z-0.01 F1000.         N219 X52.385
N122 G0 Z1.404                    N174 Z-2.96                       N220 G0 X70.
N123 X22.828                      N175 X7.                          N221 Z5.
N124 G1 X20. Z-0.01 F1000.        N176 X9.828 Z-1.546
N125 Z-2.988                      N177 G0 Z1.404                    N222 M9
N126 G3 X23. Z-3.504 R3.8         N178 X4.828                       N223 G53 X0.
N127 G1 X25.828 Z-2.09            N179 G1 X2. Z-0.01 F1000.         N224 G53 Z0.
N128 G0 Z1.404                    N180 Z-2.96                       N225 M30
N129 X20.828                      N181 X5.                          %
N130 G1 X18. Z-0.01 F1000.        N182 X7.828 Z-1.546
```

Setup Sheet - Part 9002

Job

WCS: #0

STOCK:

DX: 50mm
DY: 50mm
DZ: 120mm

PART:

DX: 48.99mm
DY: 48.99mm
DZ: 112.08mm

STOCK LOWER IN WCS #0:
X: -25mm

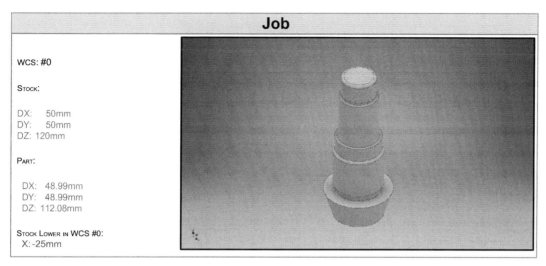

Total

NUMBER OF OPERATIONS: 1
NUMBER OF TOOLS: 1
TOOLS: T1

MAXIMUM Z: 5mm

MINIMUM Z: -101.84mm

MAXIMUM FEEDRATE: 1000mm/min
MAXIMUM SPINDLE SPEED: 500rpm

Operation 1/1

DESCRIPTION: Profile1
STRATEGY: Unspecified
WCS: #0

TOLERANCE: 0.01mm
STOCK TO LEAVE: 0mm

MAXIMUM Z: 5mm

MINIMUM Z: -101.84mm

MAXIMUM SPINDLE SPEED: 500rpm
MAXIMUM FEEDRATE: 1000mm/min
CUTTING DISTANCE: 1201.3mm
RAPID DISTANCE: 1301.61mm

T1 D0 L0

TYPE: general turning

DIAMETER: 0mm

Generated by Inventor HSM Pro 4.0.0.032

127

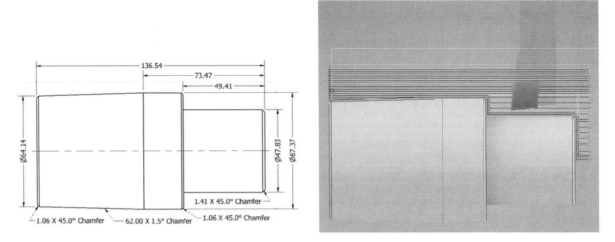

Figure 5.5 Drawing and tool path for part 0010

%
O0010 (Part 0010)
N10 G98 G18
N11 G21
N12 G50 S6000
N13 M31
N14 G53 G0 X0.

(Face2)
N15 T101
N16 G98
N17 M22
N18 G97 S500 M3
N19 G54
N20 M8
N21 G0 X121.92 Z5.
N22 G0 Z-6.495
N23 G1 X104.474 F1000.
N24 X101.6 Z-7.932
N25 X-1.6
N26 X1.274 Z-6.495
N27 G0 X121.92
N28 Z5.

(Profile2)
N29 G0 Z5.
N30 X121.92
N31 Z1.427
N32 X104.417
N33 G1 X102.442 F1000.
N34 X99.568 Z-0.01
N35 Z-144.508
N36 X101.58
N37 X104.454 Z-143.071
N38 G0 Z1.427
N39 X100.41
N40 G1 X97.536 Z-0.01 F1000.
N41 Z-144.508
N42 X99.568
N43 X102.442 Z-143.071
N44 G0 Z1.427
N45 X98.378
N46 G1 X95.504 Z-0.01 F1000.
N47 Z-144.508

N48 X97.536
N49 X100.41 Z-143.071
N50 G0 Z1.427
N51 X96.346
N52 G1 X93.472 Z-0.01 F1000.
N53 Z-144.508
N54 X95.504
N55 X98.378 Z-143.071
N56 G0 Z1.427
N57 X94.314
N58 G1 X91.44 Z-0.01 F1000.
N59 Z-144.508
N60 X93.472
N61 X96.346 Z-143.071
N62 G0 Z1.427
N63 X92.282
N64 G1 X89.408 Z-0.01 F1000.
N65 Z-144.508
N66 X91.44
N67 X94.314 Z-143.071
N68 G0 Z1.427
N69 X90.25
N70 G1 X87.376 Z-0.01 F1000.
N71 Z-144.508
N72 X89.408
N73 X92.282 Z-143.071
N74 G0 Z1.427
N75 X88.218
N76 G1 X85.344 Z-0.01 F1000.
N77 Z-144.508
N78 X87.376
N79 X90.25 Z-143.071
N80 G0 Z1.427
N81 X86.186
N82 G1 X83.312 Z-0.01 F1000.
N83 Z-144.508
N84 X85.344
N85 X88.218 Z-143.071
N86 G0 Z1.427
N87 X84.154
N88 G1 X81.28 Z-0.01 F1000.
N89 Z-144.508
N90 X83.312
N91 X86.186 Z-143.071

N92 G0 Z1.427
N93 X82.122
N94 G1 X79.248 Z-0.01 F1000.
N95 Z-144.508
N96 X81.28
N97 X84.154 Z-143.071
N98 G0 Z1.427
N99 X80.09
N100 G1 X77.216 Z-0.01 F1000.
N101 Z-144.508
N102 X79.248
N103 X82.122 Z-143.071
N104 G0 Z1.427
N105 X78.058
N106 G1 X75.184 Z-0.01 F1000.
N107 Z-144.508
N108 X77.216
N109 X80.09 Z-143.071
N110 G0 Z1.427
N111 X76.026
N112 G1 X73.152 Z-0.01 F1000.
N113 Z-144.508
N114 X75.184
N115 X78.058 Z-143.071
N116 G0 Z1.427
N117 X73.994
N118 G1 X71.12 Z-0.01 F1000.
N119 Z-144.508
N120 X73.152
N121 X76.026 Z-143.071
N122 G0 Z1.427
N123 X71.962
N124 G1 X69.088 Z-0.01 F1000.
N125 Z-57.89
N126 G18 G3 X69.406 Z-58.447 R1.056
N127 G1 Z-81.447
N128 G3 X69.405 Z-81.475 R1.056
N129 G1 X69.088 Z-87.557

128

N130 Z-144.508
N131 X71.12
N132 X73.994 Z-143.071
N133 G0 Z1.427
N134 X69.93
N135 G1 X67.056 Z-0.01
F1000.
N136 Z-56.835
N137 X68.787 Z-57.7
N138 G3 X69.406 Z-58.447
R1.056
N139 G1 X72.279 Z-57.01
N140 G0 Z1.427
N141 X67.898
N142 G1 X65.024 Z-0.01
F1000.
N143 Z-56.33
N144 X65.172
N145 G3 X66.666 Z-56.64
R1.056
N146 G1 X68.072 Z-57.343
N147 X70.946 Z-55.906
N148 G0 Z1.427
N149 X65.866
N150 G1 X62.992 Z-0.01
F1000.
N151 Z-56.33
N152 X65.172
N153 G3 X66.04 Z-56.424
R1.056
N154 G1 X68.914 Z-54.987
N155 G0 Z1.427
N156 X63.834
N157 G1 X60.96 Z-0.01 F1000.
N158 Z-56.33
N159 X64.008
N160 X66.882 Z-54.894
N161 G0 Z1.427
N162 X61.802
N163 G1 X58.928 Z-0.01
F1000.
N164 Z-56.33
N165 X61.976
N166 X64.85 Z-54.894
N167 G0 Z1.427
N168 X59.77
N169 G1 X56.896 Z-0.01
F1000.
N170 Z-56.33
N171 X59.944
N172 X62.818 Z-54.894
N173 G0 Z1.427
N174 X57.738
N175 G1 X54.864 Z-0.01
F1000.
N176 Z-56.33
N177 X57.912
N178 X60.786 Z-54.894
N179 G0 Z1.427
N180 X55.706
N181 G1 X52.832 Z-0.01
F1000.
N182 Z-56.33
N183 X55.88
N184 X58.754 Z-54.894
N185 G0 Z1.427
N186 X53.674
N187 G1 X50.8 Z-0.01 F1000.
N188 Z-56.33
N189 X53.848
N190 X56.722 Z-54.894
N191 G0 Z1.427
N192 X51.642

N193 G1 X48.768 Z-0.01
F1000.
N194 Z-8.403
N195 X49.242 Z-8.64
N196 G3 X49.86 Z-9.386
R1.056
N197 G1 Z-56.33
N198 X51.816
N199 X54.69 Z-54.894
N200 G0 Z1.427
N201 X49.61
N202 G1 X46.736 Z-0.01
F1000.
N203 Z-7.387
N204 X49.242 Z-8.64
N205 G3 X49.784 Z-9.105
R1.056
N206 G1 X52.658 Z-7.668
N207 G0 Z1.427
N208 X47.578
N209 G1 X44.704 Z-0.01
F1000.
N210 Z-6.916
N211 X44.92
N212 G3 X46.413 Z-7.226
R1.056
N213 G1 X47.752 Z-7.895
N214 X50.626 Z-6.458
N215 G0 Z1.427
N216 X45.546
N217 G1 X42.672 Z-0.01
F1000.
N218 Z-6.916
N219 X44.92
N220 G3 X45.72 Z-6.995
R1.056
N221 G1 X48.594 Z-5.558
N222 G0 Z1.427
N223 X43.514
N224 G1 X40.64 Z-0.01 F1000.
N225 Z-6.916
N226 X43.688
N227 X46.562 Z-5.479
N228 G0 Z1.427
N229 X41.482
N230 G1 X38.608 Z-0.01
F1000.
N231 Z-6.916
N232 X41.656
N233 X44.53 Z-5.479
N234 G0 Z1.427
N235 X39.45
N236 G1 X36.576 Z-0.01
F1000.
N237 Z-6.916
N238 X39.624
N239 X42.498 Z-5.479
N240 G0 Z1.427
N241 X37.418
N242 G1 X34.544 Z-0.01
F1000.
N243 Z-6.916
N244 X37.592
N245 X40.466 Z-5.479
N246 G0 Z1.427
N247 X35.386
N248 G1 X32.512 Z-0.01
F1000.
N249 Z-6.916
N250 X35.56
N251 X38.434 Z-5.479
N252 G0 Z1.427
N253 X33.354

N254 G1 X30.48 Z-0.01 F1000.
N255 Z-6.916
N256 X33.528
N257 X36.402 Z-5.479
N258 G0 Z1.427
N259 X31.322
N260 G1 X28.448 Z-0.01
F1000.
N261 Z-6.916
N262 X31.496
N263 X34.37 Z-5.479
N264 G0 Z1.427
N265 X29.29
N266 G1 X26.416 Z-0.01
F1000.
N267 Z-6.916
N268 X29.464
N269 X32.338 Z-5.479
N270 G0 Z1.427
N271 X27.258
N272 G1 X24.384 Z-0.01
F1000.
N273 Z-6.916
N274 X27.432
N275 X30.306 Z-5.479
N276 G0 Z1.427
N277 X25.226
N278 G1 X22.352 Z-0.01
F1000.
N279 Z-6.916
N280 X25.4
N281 X28.274 Z-5.479
N282 G0 Z1.427
N283 X23.194
N284 G1 X20.32 Z-0.01 F1000.
N285 Z-6.916
N286 X23.368
N287 X26.242 Z-5.479
N288 G0 Z1.427
N289 X21.162
N290 G1 X18.288 Z-0.01
F1000.
N291 Z-6.916
N292 X21.336
N293 X24.21 Z-5.479
N294 G0 Z1.427
N295 X19.13
N296 G1 X16.256 Z-0.01
F1000.
N297 Z-6.916
N298 X19.304
N299 X22.178 Z-5.479
N300 G0 Z1.427
N301 X17.098
N302 G1 X14.224 Z-0.01
F1000.
N303 Z-6.916
N304 X17.272
N305 X20.146 Z-5.479
N306 G0 Z1.427
N307 X15.066
N308 G1 X12.192 Z-0.01
F1000.
N309 Z-6.916
N310 X15.24
N311 X18.114 Z-5.479
N312 G0 Z1.427
N313 X13.034
N314 G1 X10.16 Z-0.01 F1000.
N315 Z-6.916
N316 X13.208
N317 X16.082 Z-5.479
N318 G0 Z1.427

```
N319 X11.002                      N346 X3.048                        N370  G3   X44.977   Z-7.944
N320 G1 X8.128 Z-0.01 F1000.      N347 X5.922 Z-5.479                R0.04
N321 Z-6.916                      N348 X48.563                       N371 G1 X47.805 Z-9.358
N322 X11.176                      N349 G0 X71.962                    N372  G3   X47.828   Z-9.386
N323 X14.05 Z-5.479               N350 Z-86.12                       R0.04
N324 G0 Z1.427                    N351  G1   X69.088   Z-87.557      N373 G1 Z-57.346
N325 X8.97                        F1000.                             N374 X65.172
N326 G1 X6.096 Z-0.01 F1000.      N352 X67.542 Z-117.212             N375  G3   X65.229   Z-57.358
N327 Z-6.916                      N353 Z-144.508                     R0.04
N328 X9.144                       N354 X69.088                       N376 G1 X67.35 Z-58.419
N329 X12.018 Z-5.479              N355 X71.962 Z-143.071             N377  G3   X67.374   Z-58.447
N330 G0 Z1.427                    N356 G0 Z-115.775                  R0.04
N331 X6.938                       N357 X71.682                       N378 G1 Z-81.447
N332 G1 X4.064 Z-0.01 F1000.      N358 G1 X70.415 F1000.             N379 Z-81.448
N333 Z-6.916                      N359 X67.542 Z-117.212             N380 X64.141 Z-143.448
N334 X7.112                       N360 X66.172 Z-143.475             N381 Z-143.45
N335 X9.986 Z-5.479               N361  G3   X66.165   Z-143.539     N382 X63.956 Z-144.508
N336 G0 Z1.427                    R1.056                             N383 X66.83 Z-143.071
N337 X4.906                       N362 G1 X65.996 Z-144.508          N384 X68.226
N338 G1 X2.032 Z-0.01 F1000.      N363 X67.542                       N385 G0 X121.92
N339 Z-6.916                      N364 X70.415 Z-143.071             N386 Z5.
N340 X5.08                        N365 G0 Z-6.495
N341 X7.954 Z-5.479               N366 X47.825                       N387 M9
N342 G0 Z1.427                    N367 G1 X1.279 F1000.              N388 G53 X0.
N343 X2.874                       N368 X-1.595 Z-7.932               N389 G53 Z0.
N344 G1 X0. Z-0.01 F1000.         N369 X44.92                        N390 M30
N345 Z-6.916                                                         %
```

Setup Sheet – Part 0010

Job

WCS: #0

STOCK:

DX: 101.6mm
DY: 101.6mm
DZ: 152.4mm

PART:

DX: 67.37mm
DY: 67.37mm
DZ: 136.54mm

STOCK LOWER IN WCS #0:
X: -50.8mm

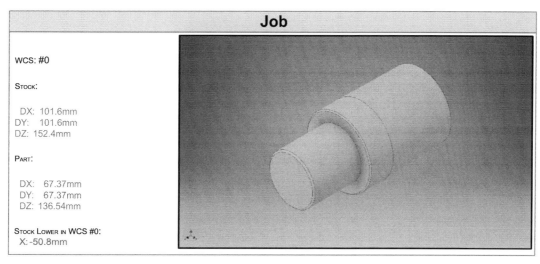

Total

NUMBER OF OPERATIONS: 2
NUMBER OF TOOLS: 1
TOOLS: T1

MAXIMUM Z: 5mm

MINIMUM Z: -144.51mm

MAXIMUM FEEDRATE: 1000mm/min
MAXIMUM SPINDLE SPEED: 500rpm

Operation 1/2

DESCRIPTION: Face2   MAXIMUM Z: 5mm   **T1** D0 L0

STRATEGY: Unspecified   MINIMUM Z: -7.93mm   TYPE: general turning
WCS: #0   MAXIMUM SPINDLE SPEED: 500rpm   DIAMETER: 0mm
TOLERANCE: 0.01mm   MAXIMUM FEEDRATE: 1000mm/min   LENGTH: 0mm

CUTTING DISTANCE: 64.39mm   FLUTES: 1

Operation 2/2

DESCRIPTION: Profile2   MAXIMUM Z: 5mm   **T1** D0 L0

STRATEGY: Unspecified   MINIMUM Z: -144.51mm   TYPE: general turning
WCS: #0   MAXIMUM SPINDLE SPEED: 500rpm   DIAMETER: 0mm
TOLERANCE: 0.01mm   MAXIMUM FEEDRATE: 1000mm/min   LENGTH: 0mm

STOCK TO LEAVE: 0mm   CUTTING DISTANCE: 3629.43mm   FLUTES: 1

Generated by Inventor HSM Pro 4.0.0.032

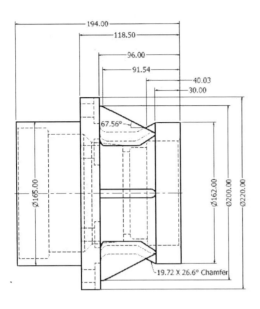
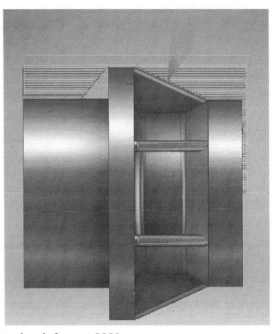

Figure 5.6 Drawing and tool path for part 9001

```
%

O9001 (Part 9001)
N10 G98 G18
N11 G21
N12 G50 S6000
N13 M31
N14 G53 G0 X0.

(Profile1)
N15 T202
N16 G98
N17 M22
N18 G97 S500 M3
N19 G54
N20 M8
N21 G0 X240. Z5.
N22 G0 Z1.404
N23 X222.581
N24 G1 X220.828 F1000.
N25 X218. Z-0.01
N26 Z-94.419
N27 X219.98 Z-96.659
N28 X222.808 Z-95.245
N29 G0 Z1.404
N30 X218.828
N31 G1 X216. Z-0.01 F1000.
N32 Z-92.156
N33 X219. Z-95.55
N34 X221.828 Z-94.136
N35 G0 Z1.404
N36 X216.828
N37 G1 X214. Z-0.01 F1000.
N38 Z-89.892
N39 X215.311 Z-91.377
N40 X217. Z-93.288
N41 X219.828 Z-91.873
N42 G0 Z1.404
```

```
N43 X214.828
N44 G1 X212. Z-0.01 F1000.
N45 Z-87.628
N46 X212.963 Z-88.718
N47 X212.964
N48 X215. Z-91.024
N49 X217.828 Z-89.61
N50 G0 Z1.404
N51 X212.828
N52 G1 X210. Z-0.01 F1000.
N53 Z-85.364
N54 X210.613 Z-86.059
N55 X212.963 Z-88.718
N56 X212.964
N57 X213. Z-88.759
N58 X215.828 Z-87.345
N59 G0 Z1.404
N60 X210.828
N61 G1 X208. Z-0.01 F1000.
N62 Z-83.1
N63 X208.265 Z-83.4
N64 X208.266
N65 X210.613 Z-86.059
N66 X211. Z-86.496
N67 X213.828 Z-85.082
N68 G0 Z1.404
N69 X208.828
N70 G1 X206. Z-0.01 F1000.
N71 Z-80.837
N72 X208.265 Z-83.4
N73 X208.266
N74 X209. Z-84.232
N75 X211.828 Z-82.818
N76 G0 Z1.404
N77 X206.828
N78 G1 X204. Z-0.01 F1000.
N79 Z-78.572
```

```
N80 X205.915 Z-80.741
N81 X207. Z-81.968
N82 X209.828 Z-80.554
N83 G0 Z1.404
N84 X204.828
N85 G1 X202. Z-0.01 F1000.
N86 Z-76.308
N87 X203.567 Z-78.082
N88 X203.568
N89 X205. Z-79.704
N90 X207.828 Z-78.29
N91 G0 Z1.404
N92 X202.828
N93 G1 X200. Z-0.01 F1000.
N94 Z-74.045
N95 X201.217 Z-75.423
N96 X203. Z-77.44
N97 X205.828 Z-76.026
N98 G0 Z1.404
N99 X200.828
N100 G1 X198. Z-0.01
F1000.
N101 Z-71.781
N102 X198.868 Z-72.764
N103 X201. Z-75.177
N104 X203.828 Z-73.763
N105 G0 Z1.404
N106 X198.828
N107 G1 X196. Z-0.01
F1000.
N108 Z-69.517
N109 X198.868 Z-72.764
N110 X199. Z-72.913
N111 X201.828 Z-71.499
N112 G0 Z1.404
N113 X196.828
N114 G1 X194. Z-0.01
```

```
F1000.                      N178 Z-46.879                    N242 G0 Z1.404
N115 Z-67.254               N179 X177.726 Z-48.833           N243 X160.828
N116 X194.17 Z-67.446       N180 X179. Z-50.275              N244 G1 X158. Z-0.01
N117 X197. Z-70.649         N181 X181.828 Z-48.861           F1000.
N118 X199.828 Z-69.235      N182 G0 Z1.404                   N245 Z-2.
N119 G0 Z1.404              N183 X176.828                    N246 X161.
N120 X194.828               N184 G1 X174. Z-0.01             N247 X163.828 Z-0.586
N121 G1 X192. Z-0.01        F1000.                           N248 G0 Z1.404
F1000.                      N185 Z-44.615                    N249 X158.828
N122 Z-64.99                N186 X175.377 Z-46.174           N250 G1 X156. Z-0.01
N123 X194.17 Z-67.446       N187 X177. Z-48.011              F1000.
N124 X195. Z-68.385         N188 X179.828 Z-46.597           N251 Z-2.
N125 X197.828 Z-66.971      N189 G0 Z1.404                   N252 X159.
N126 G0 Z1.404              N190 X174.828                    N253 X161.828 Z-0.586
N127 X192.828               N191 G1 X172. Z-0.01             N254 G0 Z1.404
N128 G1 X190. Z-0.01        F1000.                           N255 X156.828
F1000.                      N192 Z-42.352                    N256 G1 X154. Z-0.01
N129 Z-62.726               N193 X173.028 Z-43.515           F1000.
N130 X191.821 Z-64.787      N194 X175. Z-45.747              N257 Z-2.
N131 X193. Z-66.122         N195 X177. Z-44.333              N258 X157.
N132 X195.828 Z-64.707      N196 G0 Z1.404                   N259 X159.828 Z-0.586
N133 G0 Z1.404              N197 X172.828                    N260 G0 Z1.404
N134 X190.828               N198 G1 X170. Z-0.01             N261 X154.828
N135 G1 X188. Z-0.01        F1000.                           N262 G1 X152. Z-0.01
F1000.                      N199 Z-40.088                    F1000.
N136 Z-60.462               N200 X170.679 Z-40.856           N263 Z-2.
N137 X189.472 Z-62.128      N201 X173. Z-43.484              N264 X155.
N138 X191. Z-63.858         N202 X175.828 Z-42.069           N265 X157.828 Z-0.586
N139 X193.828 Z-62.443      N203 G0 Z1.404                   N266 G0 Z1.404
N140 G0 Z1.404              N204 X170.828                    N267 X152.828
N141 X188.828               N205 G1 X168. Z-0.01             N268 G1 X150. Z-0.01
N142 G1 X186. Z-0.01        F1000.                           F1000.
F1000.                      N206 Z-37.824                    N269 Z-2.
N143 Z-58.198               N207 X168.33 Z-38.197            N270 X153.
N144 X187.123 Z-59.469      N208 X170.679 Z-40.856           N271 X155.828 Z-0.586
N145 X189. Z-61.594         N209 X171. Z-41.22               N272 G0 Z1.404
N146 X191.828 Z-60.18       N210 X173.828 Z-39.805           N273 X150.828
N147 G0 Z1.404              N211 G0 Z1.404                   N274 G1 X148. Z-0.01
N148 X186.828               N212 X168.828                    F1000.
N149 G1 X184. Z-0.01        N213 G1 X166. Z-0.01             N275 Z-2.
F1000.                      F1000.                           N276 X151.
N150 Z-55.934               N214 Z-35.56                     N277 X153.828 Z-0.586
N151 X184.774 Z-56.81       N215 X168.33 Z-38.197            N278 G0 Z1.404
N152 X187. Z-59.33          N216 X169. Z-38.956              N279 X148.828
N153 X189.828 Z-57.916      N217 X171.828 Z-37.542           N280 G1 X146. Z-0.01
N154 G0 Z1.404              N218 G0 Z1.404                   F1000.
N155 X184.828               N219 X166.828                    N281 Z-2.
N156 G1 X182. Z-0.01        N220 G1 X164. Z-0.01             N282 X149.
F1000.                      F1000.                           N283 X151.828 Z-0.586
N157 Z-53.671               N221 Z-3.192                     N284 G0 Z1.404
N158 X184.774 Z-56.81       N222 Z-3.2                       N285 X146.828
N159 X185. Z-57.066         N223 Z-33.2                      N286 G1 X144. Z-0.01
N160 X187.828 Z-55.652      N224 G18 G3 X163.998 Z-          F1000.
N161 G0 Z1.404              33.244 R1.2                      N287 Z-2.
N162 X182.828               N225 G1 Z-33.295                 N288 X147.
N163 G1 X180. Z-0.01        N226 X165.981 Z-35.538           N289 X149.828 Z-0.586
F1000.                      N227 X167. Z-36.692              N290 G0 Z1.404
N164 Z-51.407               N228 X169.828 Z-35.278           N291 X144.828
N165 X180.075 Z-51.492      N229 G0 Z1.404                   N292 G1 X142. Z-0.01
N166 X183. Z-54.803         N230 X164.828                    F1000.
N167 X185.828 Z-53.388      N231 G1 X162. Z-0.01             N293 Z-2.
N168 G0 Z1.404              F1000.                           N294 X145.
N169 X180.828               N232 Z-2.017                     N295 X147.828 Z-0.586
N170 G1 X178. Z-0.01        N233 G3 X164. Z-3.2 R1.2         N296 G0 Z1.404
F1000.                      N234 G1 X166.828 Z-1.786         N297 X142.828
N171 Z-49.143               N235 G0 Z1.404                   N298 G1 X140. Z-0.01
N172 X180.075 Z-51.492      N236 X162.828                    F1000.
N173 X181. Z-52.539         N237 G1 X160. Z-0.01             N299 Z-2.
N174 X183.828 Z-51.125      F1000.                           N300 X143.
N175 G0 Z1.404              N238 Z-2.                         N301 X145.828 Z-0.586
N176 X178.828               N239 X161.6                      N302 G0 Z1.404
N177 G1 X176. Z-0.01        N240 G3 X163. Z-2.225 R1.2       N303 X140.828
F1000.                      N241 G1 X165.828 Z-0.811         N304 G1 X138. Z-0.01
```

```
F1000.                    N367 Z-2.24                  N431 X98.828
N305 Z-2.                 N368 X121. Z-2.109           N432 G1 X96. Z-0.01 F1000.
N306 X141.                N369 X123.828 Z-0.695        N433 Z-3.203
N307 X143.828 Z-0.586     N370 G0 Z1.404               N434 X99. Z-3.072
N308 G0 Z1.404            N371 X118.828                N435 X101.828 Z-1.657
N309 X138.828            N372 G1 X116. Z-0.01          N436 G0 Z1.404
N310 G1 X136. Z-0.01      F1000.                       N437 X96.828
F1000.                    N373 Z-2.328                 N438 G1 X94. Z-0.01 F1000.
N311 Z-2.                 N374 X119. Z-2.197           N439 Z-3.29
N312 X139.                N375 X121.828 Z-0.782        N440 X97. Z-3.159
N313 X141.828 Z-0.586     N376 G0 Z1.404               N441 X99.828 Z-1.745
N314 G0 Z1.404            N377 X116.828                N442 G0 Z1.404
N315 X136.828            N378 G1 X114. Z-0.01          N443 X94.828
N316 G1 X134. Z-0.01      F1000.                       N444 G1 X92. Z-0.01 F1000.
F1000.                    N379 Z-2.415                 N445 Z-3.378
N317 Z-2.                 N380 X117. Z-2.284           N446 X95. Z-3.247
N318 X137.                N381 X119.828 Z-0.87         N447 X97.828 Z-1.832
N319 X139.828 Z-0.586     N382 G0 Z1.404               N448 G0 Z1.404
N320 G0 Z1.404            N383 X114.828                N449 X92.828
N321 X134.828            N384 G1 X112. Z-0.01          N450 G1 X90. Z-0.01 F1000.
N322 G1 X132. Z-0.01      F1000.                       N451 Z-3.465
F1000.                    N385 Z-2.503                 N452 X93. Z-3.334
N323 Z-2.                 N386 X115. Z-2.372           N453 X95.828 Z-1.92
N324 X135.                N387 X117.828 Z-0.957        N454 G0 Z1.404
N325 X137.828 Z-0.586     N388 G0 Z1.404               N455 X90.828
N326 G0 Z1.404            N389 X112.828                N456 G1 X88. Z-0.01 F1000.
N327 X132.828            N390 G1 X110. Z-0.01          N457 Z-3.553
N328 G1 X130. Z-0.01      F1000.                       N458 X91. Z-3.421
F1000.                    N391 Z-2.59                  N459 X93.828 Z-2.007
N329 Z-2.                 N392 X113. Z-2.459           N460 G0 Z1.404
N330 X133.                N393 X115.828 Z-1.045        N461 X88.828
N331 X135.828 Z-0.586     N394 G0 Z1.404               N462 G1 X86. Z-0.01 F1000.
N332 G0 Z1.404            N395 X110.828                N463 Z-3.64
N333 X130.828            N396 G1 X108. Z-0.01          N464 X89. Z-3.509
N334 G1 X128. Z-0.01      F1000.                       N465 X91.828 Z-2.095
F1000.                    N397 Z-2.678                 N466 G0 Z1.404
N335 Z-2.                 N398 X111. Z-2.547           N467 X86.828
N336 X131.                N399 X113.828 Z-1.132        N468 G1 X84. Z-0.01 F1000.
N337 X133.828 Z-0.586     N400 G0 Z1.404               N469 Z-3.728
N338 G0 Z1.404            N401 X108.828                N470 X87. Z-3.596
N339 X128.828            N402 G1 X106. Z-0.01          N471 X89.828 Z-2.182
N340 G1 X126. Z-0.01      F1000.                       N472 G0 Z1.404
F1000.                    N403 Z-2.765                 N473 X84.828
N341 Z-2.                 N404 X109. Z-2.634           N474 G1 X82. Z-0.01 F1000.
N342 X129.                N405 X111.828 Z-1.22         N475 Z-3.815
N343 X131.828 Z-0.586     N406 G0 Z1.404               N476 X85. Z-3.684
N344 G0 Z1.404            N407 X106.828                N477 X87.828 Z-2.27
N345 X126.828            N408 G1 X104. Z-0.01          N478 G0 Z1.404
N346 G1 X124. Z-0.01      F1000.                       N479 X82.828
F1000.                    N409 Z-2.853                 N480 G1 X80. Z-0.01 F1000.
N347 Z-2.                 N410 X107. Z-2.722           N481 Z-3.903
N348 X127.                N411 X109.828 Z-1.307        N482 X83. Z-3.771
N349 X129.828 Z-0.586     N412 G0 Z1.404               N483 X85.828 Z-2.357
N350 G0 Z1.404            N413 X104.828                N484 G0 Z1.404
N351 X124.828            N414 G1 X102. Z-0.01          N485 X80.828
N352 G1 X122. Z-0.01      F1000.                       N486 G1 X78. Z-0.01 F1000.
F1000.                    N415 Z-2.94                  N487 Z-3.99
N353 Z-2.065              N416 X105. Z-2.809           N488 X81. Z-3.859
N354 X123.391 Z-2.005     N417 X107.828 Z-1.395        N489 X83.828 Z-2.445
N355 G3 X123.6 Z-2. R1.2  N418 G0 Z1.404               N490 G0 Z1.404
N356 G1 X125.             N419 X102.828                N491 X78.828
N357 X127.828 Z-0.586     N420 G1 X100. Z-0.01         N492 G1 X76. Z-0.01 F1000.
N358 G0 Z1.404            F1000.                       N493 Z-4.078
N359 X122.828            N421 Z-3.028                  N494 X79. Z-3.946
N360 G1 X120. Z-0.01      N422 X103. Z-2.897           N495 X81.828 Z-2.532
F1000.                    N423 X105.828 Z-1.482        N496 G0 Z1.404
N361 Z-2.153              N424 G0 Z1.404               N497 X76.828
N362 X123. Z-2.022        N425 X100.828                N498 G1 X74. Z-0.01 F1000.
N363 X125.828 Z-0.607     N426 G1 X98. Z-0.01 F1000.   N499 Z-4.165
N364 G0 Z1.404            N427 Z-3.115                  N500 X77. Z-4.034
N365 X120.828            N428 X101. Z-2.984            N501 X79.828 Z-2.62
N366 G1 X118. Z-0.01      N429 X103.828 Z-1.57         N502 G0 Z1.404
F1000.                    N430 G0 Z1.404               N503 X74.828
```

```
N504 G1 X72. Z-0.01 F1000.    N577 Z-5.302                   N650 X27. Z-6.221
N505 Z-4.253                   N578 X51. Z-5.171              N651 X29.828 Z-4.807
N506 X75. Z-4.121              N579 X53.828 Z-3.757           N652 G0 Z1.404
N507 X77.828 Z-2.707           N580 G0 Z1.404                 N653 X24.828
N508 G0 Z1.404                 N581 X48.828                   N654 G1 X22. Z-0.01 F1000.
N509 X72.828                   N582 G1 X46. Z-0.01 F1000.     N655 Z-6.44
N510 G1 X70. Z-0.01 F1000.     N583 Z-5.39                    N656 X25. Z-6.309
N511 Z-4.34                    N584 X49. Z-5.259              N657 X27.828 Z-4.894
N512 X73. Z-4.209              N585 X51.828 Z-3.845           N658 G0 Z1.404
N513 X75.828 Z-2.795           N586 G0 Z1.404                 N659 X22.828
N514 G0 Z1.404                 N587 X46.828                   N660 G1 X20. Z-0.01 F1000.
N515 X70.828                   N588 G1 X44. Z-0.01 F1000.     N661 Z-6.527
N516 G1 X68. Z-0.01 F1000.     N589 Z-5.477                   N662 X23. Z-6.396
N517 Z-4.428                   N590 X47. Z-5.346              N663 X25.828 Z-4.982
N518 X71. Z-4.296              N591 X49.828 Z-3.932           N664 G0 Z1.404
N519 X73.828 Z-2.882           N592 G0 Z1.404                 N665 X20.828
N520 G0 Z1.404                 N593 X44.828                   N666 G1 X18. Z-0.01 F1000.
N521 X68.828                   N594 G1 X42. Z-0.01 F1000.     N667 Z-6.615
N522 G1 X66. Z-0.01 F1000.     N595 Z-5.565                   N668 X21. Z-6.484
N523 Z-4.515                   N596 X45. Z-5.434              N669 X23.828 Z-5.069
N524 X69. Z-4.384              N597 X47.828 Z-4.02            N670 G0 Z1.404
N525 X71.828 Z-2.97            N598 G0 Z1.404                 N671 X18.828
N526 G0 Z1.404                 N599 X42.828                   N672 G1 X16. Z-0.01 F1000.
N527 X66.828                   N600 G1 X40. Z-0.01 F1000.     N673 Z-6.702
N528 G1 X64. Z-0.01 F1000.     N601 Z-5.652                   N674 X19. Z-6.571
N529 Z-4.603                   N602 X43. Z-5.521              N675 X21.828 Z-5.157
N530 X67. Z-4.471              N603 X45.828 Z-4.107           N676 G0 Z1.404
N531 X69.828 Z-3.057           N604 G0 Z1.404                 N677 X16.828
N532 G0 Z1.404                 N605 X40.828                   N678 G1 X14. Z-0.01 F1000.
N533 X64.828                   N606 G1 X38. Z-0.01 F1000.     N679 Z-6.79
N534 G1 X62. Z-0.01 F1000.     N607 Z-5.74                    N680 X17. Z-6.659
N535 Z-4.69                    N608 X41. Z-5.609              N681 X19.828 Z-5.244
N536 X65. Z-4.559              N609 X43.828 Z-4.194           N682 G0 Z1.404
N537 X67.828 Z-3.145           N610 G0 Z1.404                 N683 X14.828
N538 G0 Z1.404                 N611 X38.828                   N684 G1 X12. Z-0.01 F1000.
N539 X62.828                   N612 G1 X36. Z-0.01 F1000.     N685 Z-6.877
N540 G1 X60. Z-0.01 F1000.     N613 Z-5.827                   N686 X15. Z-6.746
N541 Z-4.778                   N614 X39. Z-5.696              N687 X17.828 Z-5.332
N542 X63. Z-4.646              N615 X41.828 Z-4.282           N688 G0 Z1.404
N543 X65.828 Z-3.232           N616 G0 Z1.404                 N689 X12.828
N544 G0 Z1.404                 N617 X36.828                   N690 G1 X10. Z-0.01 F1000.
N545 X60.828                   N618 G1 X34. Z-0.01 F1000.     N691 Z-6.965
N546 G1 X58. Z-0.01 F1000.     N619 Z-5.915                   N692 X13. Z-6.834
N547 Z-4.865                   N620 X37. Z-5.784              N693 X15.828 Z-5.419
N548 X61. Z-4.734              N621 X39.828 Z-4.369           N694 G0 Z1.404
N549 X63.828 Z-3.32            N622 G0 Z1.404                 N695 X10.828
N550 G0 Z1.404                 N623 X34.828                   N696 G1 X8. Z-0.01 F1000.
N551 X58.828                   N624 G1 X32. Z-0.01 F1000.     N697 Z-7.052
N552 G1 X56. Z-0.01 F1000.     N625 Z-6.002                   N698 X11. Z-6.921
N553 Z-4.953                   N626 X35. Z-5.871              N699 X13.828 Z-5.507
N554 X59. Z-4.821              N627 X37.828 Z-4.457           N700 G0 Z1.404
N555 X61.828 Z-3.407           N628 G0 Z1.404                 N701 X8.828
N556 G0 Z1.404                 N629 X32.828                   N702 G1 X6. Z-0.01 F1000.
N557 X56.828                   N630 G1 X30. Z-0.01 F1000.     N703 Z-7.14
N558 G1 X54. Z-0.01 F1000.     N631 Z-6.09                    N704 X9. Z-7.009
N559 Z-5.04                    N632 X33. Z-5.959              N705 X11.828 Z-5.594
N560 X57. Z-4.909              N633 X35.828 Z-4.544           N706 G0 Z1.404
N561 X59.828 Z-3.495           N634 G0 Z1.404                 N707 X6.828
N562 G0 Z1.404                 N635 X30.828                   N708 G1 X4. Z-0.01 F1000.
N563 X54.828                   N636 G1 X28. Z-0.01 F1000.     N709 Z-7.227
N564 G1 X52. Z-0.01 F1000.     N637 Z-6.177                   N710 X7. Z-7.096
N565 Z-5.128                   N638 X31. Z-6.046              N711 X9.828 Z-5.682
N566 X55. Z-4.996              N639 X33.828 Z-4.632           N712 G0 Z1.404
N567 X57.828 Z-3.582           N640 G0 Z1.404                 N713 X4.828
N568 G0 Z1.404                 N641 X28.828                   N714 G1 X2. Z-0.01 F1000.
N569 X52.828                   N642 G1 X26. Z-0.01 F1000.     N715 Z-7.315
N570 G1 X50. Z-0.01 F1000.     N643 Z-6.265                   N716 X5. Z-7.184
N571 Z-5.215                   N644 X29. Z-6.134              N717 X7.828 Z-5.769
N572 X53. Z-5.084              N645 X31.828 Z-4.719           N718 G0 Z1.404
N573 X55.828 Z-3.67            N646 G0 Z1.404                 N719 X2.828
N574 G0 Z1.404                 N647 X26.828                   N720 G1 X0. Z-0.01 F1000.
N575 X50.828                   N648 G1 X24. Z-0.01 F1000.     N721 Z-7.402
N576 G1 X48. Z-0.01 F1000.     N649 Z-6.352                   N722 X3. Z-7.271
```

N723 X5.828 Z-5.857
N724 G0 Z-1.9
N725 X79.886
N726 G1 X166.827 F1000.
N727 G0 Z-31.83
N728 G1 X163.998 Z-33.244
F1000.
N729 G3 X163.993 Z-33.289
R1.2
N730 G1 X164.998 Z-34.426
N731 X167.827 Z-33.012
N732 G0 Z-1.9
N733 X165.149
N734 G1 X102.834 F1000.
N735 G0 X2.428
N736 Z-6.005
N737 G1 X-0.4 Z-7.42
F1000.
N738 Z-8.424
N739 X123.565 Z-3.001
N740 G3 X123.6 Z-3. R0.2
N741 G1 X161.6
N742 G3 X162. Z-3.2 R0.2
N743 G1 Z-33.2
N744 G3 X161.862 Z-33.351
R0.2
N745 G1 X164.151 Z-35.942
N746 X166.5 Z-38.601
N747 X168.849 Z-41.26
N748 X171.198 Z-43.919
N749 X173.547 Z-46.578
N750 X175.897 Z-49.237
N751 X178.246 Z-51.896
N752 X182.944 Z-57.214
N753 X185.293 Z-59.873
N754 X187.642 Z-62.532
N755 X189.991 Z-65.191
N756 X192.34 Z-67.85
N757 X197.039 Z-73.168
N758 X199.388 Z-75.827
N759 X201.738 Z-78.486
N760 X204.086 Z-81.145
N761 X206.436 Z-83.804
N762 X208.784 Z-86.463
N763 X211.134 Z-89.122
N764 X213.482 Z-91.781
N765 X218.182 Z-97.099
N766 G3 X219.06 Z-99. R5.2
N767 G1 X219.6
N768 G3 X219.982 Z-99.14
R0.2
N769 G1 X222.81 Z-97.726
N770 G0 Z-123.107
N771 G1 X219.98 F1000.
N772 X218. Z-123.938
N773 Z-197.2
N774 X219.98
N775 X222.808 Z-195.786
N776 G0 Z-123.938
N777 G1 X218. F1000.
N778 X216. Z-124.777
N779 Z-197.2
N780 X218.
N781 X220.828 Z-195.786
N782 G0 Z-124.777
N783 G1 X216. F1000.
N784 X214. Z-125.616
N785 Z-197.2
N786 X216.
N787 X218.828 Z-195.786
N788 G0 Z-125.616
N789 G1 X214. F1000.
N790 X212. Z-126.455

N791 Z-197.2
N792 X214.
N793 X216.828 Z-195.786
N794 G0 Z-126.455
N795 G1 X212. F1000.
N796 X210. Z-127.294
N797 Z-197.2
N798 X212.
N799 X214.828 Z-195.786
N800 G0 Z-127.294
N801 G1 X210. F1000.
N802 X208. Z-128.133
N803 Z-197.2
N804 X210.
N805 X212.828 Z-195.786
N806 G0 Z-128.133
N807 G1 X208. F1000.
N808 X206. Z-128.972
N809 Z-197.2
N810 X208.
N811 X210.828 Z-195.786
N812 G0 Z-128.972
N813 G1 X206. F1000.
N814 X204. Z-129.811
N815 Z-197.2
N816 X206.
N817 X208.828 Z-195.786
N818 G0 Z-129.811
N819 G1 X204. F1000.
N820 X202. Z-130.651
N821 Z-197.2
N822 X204.
N823 X206.828 Z-195.786
N824 G0 Z-130.651
N825 G1 X202. F1000.
N826 X200. Z-131.49
N827 Z-197.2
N828 X202.
N829 X204.828 Z-195.786
N830 G0 Z-131.49
N831 G1 X200. F1000.
N832 X198. Z-132.329
N833 Z-197.2
N834 X200.
N835 X202.828 Z-195.786
N836 G0 Z-132.329
N837 G1 X198. F1000.
N838 X196. Z-133.168
N839 Z-197.2
N840 X198.
N841 X200.828 Z-195.786
N842 G0 Z-133.168
N843 G1 X196. F1000.
N844 X194. Z-134.007
N845 Z-197.2
N846 X196.
N847 X198.828 Z-195.786
N848 G0 Z-134.007
N849 G1 X194. F1000.
N850 X192. Z-134.846
N851 Z-197.2
N852 X194.
N853 X196.828 Z-195.786
N854 G0 Z-134.846
N855 G1 X192. F1000.
N856 X190. Z-135.685
N857 Z-197.2
N858 X192.
N859 X194.828 Z-195.786
N860 G0 Z-135.685
N861 G1 X190. F1000.
N862 X188. Z-136.524
N863 Z-197.2

N864 X190.
N865 X192.828 Z-195.786
N866 G0 Z-136.524
N867 G1 X188. F1000.
N868 X186. Z-137.363
N869 Z-197.2
N870 X188.
N871 X190.828 Z-195.786
N872 G0 Z-137.363
N873 G1 X186. F1000.
N874 X184. Z-138.202
N875 Z-197.2
N876 X186.
N877 X188.828 Z-195.786
N878 G0 Z-138.202
N879 G1 X184. F1000.
N880 X182. Z-139.042
N881 Z-197.2
N882 X184.
N883 X186.828 Z-195.786
N884 G0 Z-139.042
N885 G1 X182. F1000.
N886 X180. Z-139.881
N887 Z-197.2
N888 X182.
N889 X184.828 Z-195.786
N890 G0 Z-139.881
N891 G1 X180. F1000.
N892 X178. Z-140.72
N893 Z-197.2
N894 X180.
N895 X182.828 Z-195.786
N896 G0 Z-140.72
N897 G1 X178. F1000.
N898 X176. Z-141.559
N899 Z-197.2
N900 X178.
N901 X180.828 Z-195.786
N902 G0 Z-141.559
N903 G1 X176. F1000.
N904 X174. Z-142.398
N905 Z-197.2
N906 X176.
N907 X178.828 Z-195.786
N908 G0 Z-142.398
N909 G1 X174. F1000.
N910 X172. Z-143.237
N911 Z-197.2
N912 X174.
N913 X176.828 Z-195.786
N914 G0 Z-143.237
N915 G1 X172. F1000.
N916 X170. Z-144.076
N917 Z-197.2
N918 X172.
N919 X174.828 Z-195.786
N920 G0 Z-144.076
N921 G1 X170. F1000.
N922 X168.5 Z-144.705
N923 Z-197.2
N924 X170.
N925 X172.828 Z-195.786
N926 G0 Z-144.705
N927 G1 X168.5 F1000.
N928 X167. Z-145.335
N929 Z-197.2
N930 X168.5
N931 X171.328 Z-195.786
N932 G0 X222.809
N933 Z-120.347
N934 G1 X219.981 Z-121.761
F1000.
N935 G3 X219.857 Z-121.853

```
R0.2                              N940 G0 X240.                    N944 G53 Z0.
N936 G1 X165. Z-144.868           N941 Z5.                         N945 M30
N937 Z-197.2                                                       %
N938 X167.828 Z-195.786           N942 M9
N939 X169.                        N943 G53 X0.
```

Setup Sheet – Part 9003

Job

WCS: #0

STOCK:

  DX: 220mm
DY: 220mm
DZ: 200mm

PART:

  DX: 220mm
DY: 220mm
DZ: 194mm

STOCK LOWER IN WCS #0:
  X: -110mm

Total

NUMBER OF OPERATIONS: 1
NUMBER OF TOOLS: 1
TOOLS: T2

MAXIMUM Z: 5mm

MINIMUM Z: -197.2mm

MAXIMUM FEEDRATE: 1000mm/min
MAXIMUM SPINDLE SPEED: 500rpm

Operation 1/1

DESCRIPTION: Profile1
STRATEGY: Unspecified
WCS: #0

TOLERANCE: 0.01mm
STOCK TO LEAVE: 0mm

MAXIMUM Z: 5mm

MINIMUM Z: -197.2mm

MAXIMUM SPINDLE SPEED: 500rpm
MAXIMUM FEEDRATE: 1000mm/min
CUTTING DISTANCE: 5012.41mm
RAPID DISTANCE: 4600.3mm

T2 D0 L0

TYPE: general turning

DIAMETER: 0mm

Generated by Inventor HSM Pro 4.0.0.032

137
```
```

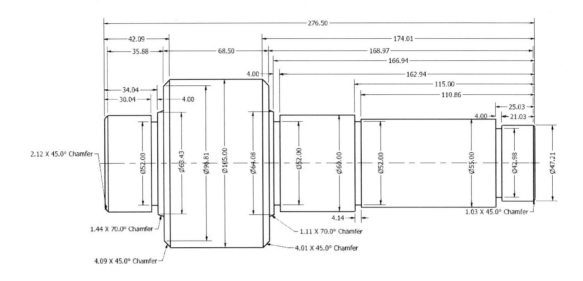

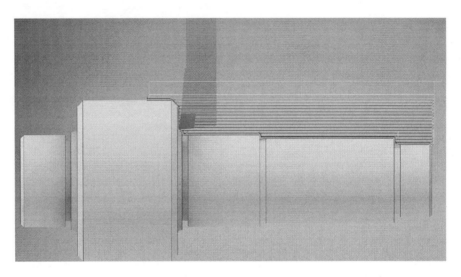

Figure 5.7 Drawing and tool path for part 0001

```
%
O0001 (Part 0001)
N10 G98 G18
N11 G21
N12 G50 S6000
N13 G28 U0.

(FACE1)
N14 T0101
N15 G54
N16 M8
N17 G98
N18 G97 S500 M3
N19 G0 X130. Z5.
N20 G0 Z1.414
N21 G1 X112.828 F1000.
N22 X110. Z0.
N23 X-1.6
```

```
N24 X1.228 Z1.414
N25 G0 X130.
N26 Z5.
N27 G28 U0.

(PROFILE2)
N28 M1
N29 T0303
N30 G54
N31 G98
N32 G97 S500 M3
N33 G0 X130. Z6.2
N34 G0 Z2.587
N35 X111.332
N36 G1 X110.828 F1000.
N37 X108. Z1.173
N38 Z-276.5
N39 X109.98
```

```
N40 X112.808 Z-275.086
N41 G0 Z2.604
N42 X108.828
N43 G1 X106. Z1.19
F1000.
N44 Z-40.698
N45 G18 G3 X107. Z-
42.094 R2.2
N46 G1 Z-102.489
N47 G3 X106. Z-103.885
R2.2
N48 G1 Z-276.5
N49 X108.
N50 X110.828 Z-275.086
N51 G0 Z2.604
N52 X106.828
N53 G1 X104. Z1.19
F1000.
```

138

```
N54 Z-39.683
N55 X105.711 Z-40.539
N56 G3 X107. Z-42.091
R2.2
N57 G1 X109.828 Z-40.677
N58 G0 Z2.604
N59 X104.828
N60 G1 X102. Z1.19
F1000.
N61 Z-38.683
N62 X105. Z-40.183
N63 X107.828 Z-38.769
N64 G0 Z2.604
N65 X102.828
N66 G1 X100. Z1.19
F1000.
N67 Z-37.683
N68 X103. Z-39.183
N69 X105.828 Z-37.769
N70 G0 Z2.604
N71 X100.828
N72 G1 X98. Z1.19 F1000.
N73 Z-36.683
N74 X101. Z-38.183
N75 X103.828 Z-36.769
N76 G0 Z2.604
N77 X98.828
N78 G1 X96. Z1.19 F1000.
N79 Z-35.918
N80 X96.612 Z-36.095
N81 G3 X97.523 Z-36.444
R2.2
N82 G1 X99. Z-37.183
N83 X101.828 Z-35.769
N84 G0 Z2.604
N85 X96.828
N86 G1 X94. Z1.19 F1000.
N87 Z-35.341
N88 X96.612 Z-36.095
N89 G3 X97. Z-36.221
R2.2
N90 G1 X99.828 Z-34.807
N91 G0 Z2.604
N92 X94.828
N93 G1 X92. Z1.19 F1000.
N94 Z-34.763
N95 X95. Z-35.63
N96 X97.828 Z-34.215
N97 G0 Z2.604
N98 X92.828
N99 G1 X90. Z1.19 F1000.
N100 Z-34.186
N101 X93. Z-35.052
N102 X95.828 Z-33.638
N103 G0 Z2.604
N104 X90.828
N105 G1 X88. Z1.19
F1000.
N106 Z-33.609
N107 X91. Z-34.475
N108 X93.828 Z-33.061
N109 G0 Z2.604
N110 X88.828
N111 G1 X86. Z1.19
F1000.
N112 Z-33.031
N113 X89. Z-33.897
N114 X91.828 Z-32.483
N115 G0 Z2.604
N116 X86.828
N117 G1 X84. Z1.19
F1000.
N118 Z-32.454

N119 X87. Z-33.32
N120 X89.828 Z-31.906
N121 G0 Z2.604
N122 X84.828
N123 G1 X82. Z1.19
F1000.
N124 Z-31.877
N125 X85. Z-32.743
N126 X87.828 Z-31.329
N127 G0 Z2.604
N128 X82.828
N129 G1 X80. Z1.19
F1000.
N130 Z-31.299
N131 X83. Z-32.165
N132 X85.828 Z-30.751
N133 G0 Z2.604
N134 X80.828
N135 G1 X78. Z1.19
F1000.
N136 Z-30.722
N137 X81. Z-31.588
N138 X83.828 Z-30.174
N139 G0 Z2.604
N140 X78.828
N141 G1 X76. Z1.19
F1000.
N142 Z-30.145
N143 X79. Z-31.011
N144 X81.828 Z-29.596
N145 G0 Z2.604
N146 X76.828
N147 G1 X74. Z1.19
F1000.
N148 Z-29.567
N149 X77. Z-30.433
N150 X79.828 Z-29.019
N151 G0 Z2.604
N152 X74.828
N153 G1 X72. Z1.19
F1000.
N154 Z-28.99
N155 X75. Z-29.856
N156 X77.828 Z-28.442
N157 G0 Z2.604
N158 X72.828
N159 G1 X70. Z1.19
F1000.
N160 Z-28.413
N161 X73. Z-29.279
N162 X75.828 Z-27.864
N163 G0 Z2.604
N164 X70.828
N165 G1 X68. Z1.19
F1000.
N166 Z-27.835
N167 X71. Z-28.701
N168 X73.828 Z-27.287
N169 G0 Z2.604
N170 X68.828
N171 G1 X66. Z1.19
F1000.
N172 Z-27.258
N173 X69. Z-28.124
N174 X71.828 Z-26.71
N175 G0 Z2.604
N176 X66.828
N177 G1 X64. Z1.19
F1000.
N178 Z-26.681
N179 X67. Z-27.547
N180 X69.828 Z-26.132
N181 G0 Z2.604

N182 X64.828
N183 G1 X62. Z1.19
F1000.
N184 Z-2.111
N185 G3 Z-2.121 R2.2
N186 G1 Z-26.103
N187 X65. Z-26.969
N188 X67.828 Z-25.555
N189 G0 Z2.604
N190 X62.828
N191 G1 X60. Z1.19
F1000.
N192 Z-0.21
N193 X60.711 Z-0.566
N194 G3 X62. Z-2.121
R2.2
N195 G1 X64.828 Z-0.707
N196 G0 Z2.604
N197 X61.418
N198 G1 X58.59 Z1.19
F1000.
N199 Z0.495
N200 X60.711 Z-0.566
N201 G3 X61. Z-0.725
R2.2
N202 G1 X63.828 Z0.689
N203 G0 Z2.604
N204 X60.029
N205 G1 X57.2 Z1.19
F1000.
N206 X59.59 Z-0.005
N207 X62.418 Z1.409
N208 G0 X108.828
N209 Z-102.471
N210 G1 X106. Z-103.885
F1000.
N211 G3 X105.711 Z-
104.045 R2.2
N212 G1 X104. Z-104.9
N213 Z-276.5
N214 X106.
N215 X108.828 Z-275.086
N216 G0 Z-104.9
N217 G1 X104. F1000.
N218 X102. Z-105.9
N219 Z-276.5
N220 X104.
N221 X106.828 Z-275.086
N222 G0 Z-105.9
N223 G1 X102. F1000.
N224 X100. Z-106.9
N225 Z-276.5
N226 X102.
N227 X104.828 Z-275.086
N228 G0 Z-106.9
N229 G1 X100. F1000.
N230 X98. Z-107.9
N231 Z-276.5
N232 X100.
N233 X102.828 Z-275.086
N234 G0 Z-107.9
N235 G1 X98. F1000.
N236 X97.689 Z-108.056
N237 G3 X96.778 Z-
108.405 R2.2
N238 G1 X96. Z-108.63
N239 Z-276.5
N240 X98.
N241 X100.828 Z-275.086
N242 G0 Z-108.63
N243 G1 X96. F1000.
N244 X94. Z-109.207
N245 Z-276.5
```

```
N246 X96.                          N319 X74.828 Z-275.086          N386 X50.626 Z-254.582
N247 X98.828 Z-275.086             N320 G0 Z-116.135               N387 Z-276.5
N248 G0 Z-109.207                  N321 G1 X70. F1000.             N388 X52.134
N249 G1 X94. F1000.                N322 X68. Z-116.713             N389 X54.962 Z-275.086
N250 X92. Z-109.784                N323 Z-276.5                    N390 G0 Z-254.582
N251 Z-276.5                       N324 X70.                       N391 G1 X50.626 F1000.
N252 X94.                          N325 X72.828 Z-275.086          N392 X49.119 Z-255.017
N253 X96.828 Z-275.086             N326 G0 Z-116.713               N393 G3 X49.214 Z-
N254 G0 Z-109.784                  N327 G1 X68. F1000.             255.472 R2.2
N255 G1 X92. F1000.                N328 X66. Z-117.29              N394 G1 Z-275.472
N256 X90. Z-110.362                N329 Z-276.5                    N395 G3 X48.704 Z-276.5
N257 Z-276.5                       N330 X68.                       R2.2
N258 X92.                          N331 X70.828 Z-275.086          N396 G1 X50.626
N259 X94.828 Z-275.086             N332 G0 Z-117.29                N397 X53.455 Z-275.086
N260 G0 Z-110.362                  N333 G1 X66. F1000.             N398 G0 X107.2
N261 G1 X90. F1000.                N334 X64. Z-117.867             N399 Z2.263
N262 X88. Z-110.939                N335 Z-276.5                    N400 X57.883
N263 Z-276.5                       N336 X66.                       N401 G1 X55.054 Z0.849
N264 X90.                          N337 X68.828 Z-275.086          F1000.
N265 X92.828 Z-275.086             N338 G0 Z-117.867               N402 X59.297 Z-1.273
N266 G0 Z-110.939                  N339 G1 X64. F1000.             N403 G3 X60. Z-2.121
N267 G1 X88. F1000.                N340 X62. Z-118.445             R1.2
N268 X86. Z-111.517                N341 Z-161.5                    N404 G1 Z-26.681
N269 Z-276.5                       N342 G3 Z-161.51 R2.2           N405 X95.612 Z-36.961
N270 X88.                          N343 G1 Z-276.5                 N406 G3 X96.109 Z-37.151
N271 X90.828 Z-275.086             N344 X64.                       R1.2
N272 G0 Z-111.517                  N345 X66.828 Z-275.086          N407 G1 X104.297 Z-
N273 G1 X86. F1000.                N346 G0 Z-160.096               41.246
N274 X84. Z-112.094                N347 X66.                       N408 G3 X105. Z-42.094
N275 Z-276.5                       N348 G1 X64.828 F1000.          R1.2
N276 X86.                          N349 X62. Z-161.51              N409 G1 Z-102.489
N277 X88.828 Z-275.086             N350 G3 X60. Z-163.344          N410 G3 X104.297 Z-
N278 G0 Z-112.094                  R2.2                            103.338 R1.2
N279 G1 X84. F1000.                N351 G1 Z-276.5                 N411 G1 X96.275 Z-
N280 X82. Z-112.671                N352 X62.                       107.349
N281 Z-276.5                       N353 X64.828 Z-275.086          N412 G3 X95.778 Z-
N282 X84.                          N354 G0 Z-163.344               107.539 R1.2
N283 X86.828 Z-275.086             N355 G1 X60. F1000.             N413 G1 X60. Z-117.867
N284 G0 Z-112.671                  N356 G3 X59.8 Z-163.405         N414 Z-161.5
N285 G1 X82. F1000.                R2.2                            N415 G3 X58.8 Z-162.539
N286 X80. Z-113.249                N357 G1 X58.134 Z-              R1.2
N287 Z-276.5                       163.886                         N416 G1 X52.729 Z-
N288 X82.                          N358 Z-276.5                    164.292
N289 X84.828 Z-275.086             N359 X60.                       N417 X53.8 Z-164.601
N290 G0 Z-113.249                  N360 X62.828 Z-275.086          N418 G3 X55. Z-165.64
N291 G1 X80. F1000.                N361 G0 Z-163.886               R1.2
N292 X78. Z-113.826                N362 G1 X58.134 F1000.          N419 G1 Z-251.472
N293 Z-276.5                       N363 X56.268 Z-164.425          N420 G3 X53.8 Z-252.511
N294 X80.                          N364 G3 X57. Z-165.64           R1.2
N295 X82.828 Z-275.086             R2.2                            N421 G1 X46.501 Z-
N296 G0 Z-113.826                  N365 G1 Z-251.472               254.618
N297 G1 X78. F1000.                N366 G3 X56.134 Z-              N422 G3 X47.214 Z-
N298 X76. Z-114.403                252.783 R2.2                    255.472 R1.2
N299 Z-276.5                       N367 G1 Z-276.5                 N423 G1 Z-275.472
N300 X78.                          N368 X58.134                    N424 G3 X46.511 Z-
N301 X80.828 Z-275.086             N369 X60.962 Z-275.086          276.321 R1.2
N302 G0 Z-114.403                  N370 G0 Z-251.368               N425 G1 X46.152 Z-276.5
N303 G1 X76. F1000.                N371 G1 X58.962 F1000.          N426 X50.873
N304 X74. Z-114.981                N372 X56.134 Z-252.783          N427 G0 X130.
N305 Z-276.5                       N373 G3 X54.8 Z-253.377         N428 Z6.2
N306 X76.                          R2.2                            N429 G28 U0.
N307 X78.828 Z-275.086             N374 G1 X54.134 Z-253.57
N308 G0 Z-114.981                  N375 Z-276.5                    (GROOVE2)
N309 G1 X74. F1000.                N376 X56.134                    N430 M1
N310 X72. Z-115.558                N377 X58.962 Z-275.086          N431 T0202
N311 Z-276.5                       N378 G0 Z-253.57                N432 G54
N312 X74.                          N379 G1 X54.134 F1000.          N433 G98
N313 X76.828 Z-275.086             N380 X52.134 Z-254.147          N434 G97 S500 M3
N314 G0 Z-115.558                  N381 Z-276.5                    N435 G0 X130. Z5.
N315 G1 X72. F1000.                N382 X54.134                    N436 G0 Z-253.472
N316 X70. Z-116.135                N383 X56.962 Z-275.086          N437 X114.
N317 Z-276.5                       N384 G0 Z-254.147               N438 G1 X110. F1000.
N318 X72.                          N385 G1 X52.134 F1000.          N439 X56.625
```

```
N440 X114.                      N505 G98                         N577 G1 X63.806 Z-28.642
N441 G0 Z-254.472               N506 G97 S500 M3                 N578 X67.216
N442 G1 X110. F1000.            N507 G0 X130. Z8.                N579 G0 X101.885
N443 X45.614                    N508 G0 Z-25.042                 N580 Z-35.
N444 X113.462                   N509 X114.                       N581 G1 X101.883 F1000.
N445 G0 X114.                   N510 G1 X110. F1000.             N582 X97.883
N446 Z-255.472                  N511 X62.                        N583 X97.347 Z-34.732
N447 G1 X110. F1000.            N512 X114.                       N584 G2 X93.812 Z-34.
N448 X48.839                    N513 G0 Z-26.037                 R2.5
N449 X56.723                    N514 G1 X110. F1000.             N585 G1 X66.812
N450 X58.723 Z-254.472          N515 X62.                        N586 X65.128 Z-31.686
N451 X110.38                    N516 X113.469                    N587 G2 X60.977 Z-30.057
N452 G0 Z-252.058               N517 G0 X114.                    R2.5
N453 G1 X59.453 F1000.          N518 Z-27.033                    N588 G1 X64.958 Z-30.253
N454 X56.625 Z-253.472          N519 G1 X110. F1000.             N589 X66.442
N455 G18 G3 X53.4 Z-            N520 X62.                        N590 G0 Z-25.042
254.472 R1.8                    N521 X64. Z-26.033               N591 G1 X64. F1000.
N456 G1 X45.614                 N522 X113.472                    N592 X60.
N457 X48.839                    N523 G0 X114.                    N593 Z-28.542
N458 Z-255.472                  N524 Z-28.029                    N594 G3 X57. Z-30.042
N459 G2 X45.614 Z-              N525 G1 X110. F1000.             R1.5
254.472 R1.8                    N526 X62.                        N595 G1 X52.
N460 G1 X49.614                 N527 X64. Z-27.029               N596 Z-31.042
N461 G0 X130.                   N528 X110.431                    N597 X60.429
N462 Z5.                        N529 G0 X114.                    N598 G3 X63.248 Z-32.028
                               N530 Z-29.025                    R1.5
(GROOVE4)                       N531 G1 X110. F1000.             N599 G1 X65.412 Z-35.
N463 G98                        N532 X61.906                     N600 X68.24 Z-33.586
N464 G97 S500 M3                N533 X62.094                     N601 X68.446
N465 G0 X130. Z5.               N534 X64.094 Z-28.025            N602 G0 X130.
N466 Z-163.5                    N535 X110.2                      N603 Z8.
N467 X114.                      N536 G0 X114.
N468 G1 X110. F1000.            N537 Z-30.021                    (GROOVE7)
N469 X61.625                    N538 G1 X110. F1000.             N604 G98
N470 X114.                      N539 X61.031                     N605 G97 S500 M3
N471 G0 Z-164.213               N540 X61.987                     N606 G0 X130. Z8.
N472 G1 X110. F1000.            N541 X63.987 Z-29.021            N607 Z-106.5
N473 X60.349                    N542 X110.255                    N608 X114.
N474 X113.737                   N543 G0 X114.                    N609 G1 X110. F1000.
N475 G0 X114.                   N544 Z-31.017                    N610 X98.049
N476 Z-164.927                  N545 G1 X110. F1000.             N611 X114.
N477 G1 X110. F1000.            N546 X64.392                     N612 G0 Z-107.5
N478 X55.349                    N547 X66.392 Z-30.017            N613 G1 X110. F1000.
N479 X112.801                   N548 X110.255                    N614 X66.812
N480 G0 X114.                   N549 G0 X114.                    N615 X113.464
N481 Z-165.64                   N550 Z-32.012                    N616 G0 X114.
N482 G1 X110. F1000.            N551 G1 X110. F1000.             N617 Z-108.5
N483 X56.625                    N552 X65.365                     N618 G1 X110. F1000.
N484 X61.42                     N553 X67.365 Z-31.012            N619 X66.084
N485 X63.42 Z-164.64            N554 X110.255                    N620 X98.163
N486 X110.38                    N555 G0 X114.                    N621 X100.163 Z-107.5
N487 G0 Z-162.086               N556 Z-33.008                    N622 X113.462
N488 G1 X64.453 F1000.          N557 G1 X110. F1000.             N623 G0 X114.
N489 X61.625 Z-163.5            N558 X66.09                      N624 Z-109.5
N490 G3 X58.4 Z-164.5           N559 X68.09 Z-32.008             N625 G1 X110. F1000.
R1.8                            N560 X110.255                    N626 X65.17
N491 G1 X54.                    N561 G0 X114.                    N627 X66.921
N492 Z-164.583                  N562 Z-34.004                    N628 X68.921 Z-108.5
N493 X56.625                    N563 G1 X110. F1000.             N629 X101.513
N494 Z-165.64                   N564 X94.1                       N630 G0 X114.
N495 G2 X54. Z-164.665          N565 X96.1 Z-33.004              N631 Z-110.5
R1.8                            N566 X101.345                    N632 G1 X110. F1000.
N496 G1 Z-164.583               N567 G0 X114.                    N633 X62.164
N497 X57.616 Z-165.438          N568 Z-35.                       N634 X66.088
N498 X58.195                    N569 G1 X110. F1000.             N635 X68.088 Z-109.5
N499 G0 X130.                   N570 X97.883                     N636 X70.276
N500 Z5.                        N571 X99.883 Z-34.               N637 G0 X114.
N501 G28 U0.                    N572 G0 Z-25.042                 N638 Z-111.5
                               N573 G1 X66. F1000.              N639 G1 X110. F1000.
(GROOVE6)                       N574 X62.                        N640 X61.873
N502 M1                         N575 Z-28.542                    N641 X65.257
N503 T0404                      N576 G18 G3 X60.977 Z-           N642 X67.257 Z-110.5
N504 G54                        30.057 R2.5                      N643 X110.
```

```
N644 G0 X114.          N717 G0 X114.          N790 Z-136.5
N645 Z-112.5           N718 Z-124.5           N791 G1 X110. F1000.
N646 G1 X110. F1000.   N719 G1 X110. F1000.   N792 X62.
N647 X62.              N720 X62.              N793 X64. Z-135.5
N648 X62.188           N721 X64. Z-123.5      N794 X110.
N649 X64.188 Z-111.5   N722 X110.             N795 G0 X114.
N650 X110.             N723 G0 X114.          N796 Z-137.5
N651 G0 X114.          N724 Z-125.5           N797 G1 X110. F1000.
N652 Z-113.5           N725 G1 X110. F1000.   N798 X62.
N653 G1 X110. F1000.   N726 X62.              N799 X64. Z-136.5
N654 X62.              N727 X64. Z-124.5      N800 X110.
N655 X64. Z-112.5      N728 X110.             N801 G0 X114.
N656 X110.             N729 G0 X114.          N802 Z-138.5
N657 G0 X114.          N730 Z-126.5           N803 G1 X110. F1000.
N658 Z-114.5           N731 G1 X110. F1000.   N804 X62.
N659 G1 X110. F1000.   N732 X62.              N805 X64. Z-137.5
N660 X62.              N733 X64. Z-125.5      N806 X110.
N661 X64. Z-113.5      N734 X110.             N807 G0 X114.
N662 X110.             N735 G0 X114.          N808 Z-139.5
N663 G0 X114.          N736 Z-127.5           N809 G1 X110. F1000.
N664 Z-115.5           N737 G1 X110. F1000.   N810 X62.
N665 G1 X110. F1000.   N738 X62.              N811 X64. Z-138.5
N666 X62.              N739 X64. Z-126.5      N812 X110.
N667 X64. Z-114.5      N740 X110.             N813 G0 X114.
N668 X110.             N741 G0 X114.          N814 Z-140.5
N669 G0 X114.          N742 Z-128.5           N815 G1 X110. F1000.
N670 Z-116.5           N743 G1 X110. F1000.   N816 X62.
N671 G1 X110. F1000.   N744 X62.              N817 X64. Z-139.5
N672 X62.              N745 X64. Z-127.5      N818 X110.
N673 X64. Z-115.5      N746 X110.             N819 G0 X114.
N674 X110.             N747 G0 X114.          N820 Z-141.5
N675 G0 X114.          N748 Z-129.5           N821 G1 X110. F1000.
N676 Z-117.5           N749 G1 X110. F1000.   N822 X62.
N677 G1 X110. F1000.   N750 X62.              N823 X64. Z-140.5
N678 X62.              N751 X64. Z-128.5      N824 X110.
N679 X64. Z-116.5      N752 X110.             N825 G0 X114.
N680 X110.             N753 G0 X114.          N826 Z-142.5
N681 G0 X114.          N754 Z-130.5           N827 G1 X110. F1000.
N682 Z-118.5           N755 G1 X110. F1000.   N828 X62.
N683 G1 X110. F1000.   N756 X62.              N829 X64. Z-141.5
N684 X62.              N757 X64. Z-129.5      N830 X110.
N685 X64. Z-117.5      N758 X110.             N831 G0 X114.
N686 X110.             N759 G0 X114.          N832 Z-143.5
N687 G0 X114.          N760 Z-131.5           N833 G1 X110. F1000.
N688 Z-119.5           N761 G1 X110. F1000.   N834 X62.
N689 G1 X110. F1000.   N762 X62.              N835 X64. Z-142.5
N690 X62.              N763 X64. Z-130.5      N836 X110.
N691 X64. Z-118.5      N764 X110.             N837 G0 X114.
N692 X110.             N765 G0 X114.          N838 Z-144.5
N693 G0 X114.          N766 Z-132.5           N839 G1 X110. F1000.
N694 Z-120.5           N767 G1 X110. F1000.   N840 X62.
N695 G1 X110. F1000.   N768 X62.              N841 X64. Z-143.5
N696 X62.              N769 X64. Z-131.5      N842 X110.
N697 X64. Z-119.5      N770 X110.             N843 G0 X114.
N698 X110.             N771 G0 X114.          N844 Z-145.5
N699 G0 X114.          N772 Z-133.5           N845 G1 X110. F1000.
N700 Z-121.5           N773 G1 X110. F1000.   N846 X62.
N701 G1 X110. F1000.   N774 X62.              N847 X64. Z-144.5
N702 X62.              N775 X64. Z-132.5      N848 X110.
N703 X64. Z-120.5      N776 X110.             N849 G0 X114.
N704 X110.             N777 G0 X114.          N850 Z-146.5
N705 G0 X114.          N778 Z-134.5           N851 G1 X110. F1000.
N706 Z-122.5           N779 G1 X110. F1000.   N852 X62.
N707 G1 X110. F1000.   N780 X62.              N853 X64. Z-145.5
N708 X62.              N781 X64. Z-133.5      N854 X110.
N709 X64. Z-121.5      N782 X110.             N855 G0 X114.
N710 X110.             N783 G0 X114.          N856 Z-147.5
N711 G0 X114.          N784 Z-135.5           N857 G1 X110. F1000.
N712 Z-123.5           N785 G1 X110. F1000.   N858 X62.
N713 G1 X110. F1000.   N786 X62.              N859 X64. Z-146.5
N714 X62.              N787 X64. Z-134.5      N860 X110.
N715 X64. Z-122.5      N788 X110.             N861 G0 X114.
N716 X110.             N789 G0 X114.          N862 Z-148.5
```

```
N863 G1 X110. F1000.       N905 G1 X110. F1000.       111.973
N864 X62.                  N906 X62.                  N944 X65.998
N865 X64. Z-147.5          N907 X64. Z-154.5          N945 G0 X102.051
N866 X110.                 N908 X110.                 N946 Z-106.5
N867 G0 X114.              N909 G0 X114.              N947 G1 X69.412 F1000.
N868 Z-149.5               N910 Z-156.5               N948 X65.412
N869 G1 X110. F1000.       N911 G1 X110. F1000.       N949 X63.903 Z-108.572
N870 X62.                  N912 X62.                  N950 G3 X61.084 Z-
N871 X64. Z-148.5          N913 X64. Z-155.5          109.559 R1.5
N872 X110.                 N914 X110.                 N951 G1 X52.
N873 G0 X114.              N915 G0 X114.              N952 Z-110.559
N874 Z-150.5               N916 Z-157.5               N953 X57.
N875 G1 X110. F1000.       N917 G1 X110. F1000.       N954 G3 X60. Z-112.059
N876 X62.                  N918 X62.                  R1.5
N877 X64. Z-149.5          N919 X64. Z-156.5          N955 G1 Z-158.5
N878 X110.                 N920 X66.                  N956 X62.828 Z-157.086
N879 G0 X114.              N921 G0 X114.              N957 X64.
N880 Z-151.5               N922 Z-158.5               N958 G0 X130.
N881 G1 X110. F1000.       N923 G1 X110. F1000.       N959 Z8.
N882 X62.                  N924 X62.                  N960 G28 U0.
N883 X64. Z-150.5          N925 X64. Z-157.5
N884 X110.                 N926 X102.051              (FACE3)
N885 G0 X114.              N927 G0 Z-106.5            N961 M1
N886 Z-152.5               N928 G1 X102.049 F1000.    N962 T0101
N887 G1 X110. F1000.       N929 X98.049               N963 G54
N888 X62.                  N930 X97.513 Z-106.768     N964 G98
N889 X64. Z-151.5          N931 G3 X93.978 Z-107.5    N965 G97 S500 M3
N890 X110.                 R2.5                       N966 G0 X171.321 Z5.
N891 G0 X114.              N932 G1 X66.812            N967 G0 Z0.414
N892 Z-153.5               N933 X65.783 Z-108.914     N968 G1 X154.149 F1000.
N893 G1 X110. F1000.       N934 G3 X61.084 Z-         N969 X151.321 Z-1.
N894 X62.                  110.559 R2.5               N970 X-1.6
N895 X64. Z-152.5          N935 G1 X61.               N971 X1.228 Z0.414
N896 X110.                 N936 X65.                  N972 G0 X171.321
N897 G0 X114.              N937 X66.166               N973 Z5.
N898 Z-154.5               N938 G0 Z-158.5
N899 G1 X110. F1000.       N939 G1 X66. F1000.        N974 M9
N900 X62.                  N940 X62.                  N975 G28 U0. W0.
N901 X64. Z-153.5          N941 Z-112.059             N976 M30
N902 X110.                 N942 G2 X61. Z-110.559     %
N903 G0 X114.              R2.5
N904 Z-155.5               N943 G1 X63.828 Z-
```

Setup Sheet – Part 0001

Job

WCS: #0

STOCK:

DX: 110mm
DY: 110mm
DZ: 280mm

PART:

DX: 105mm
DY: 105mm
DZ: 276.5mm

STOCK LOWER IN WCS #0:
X: -55mm

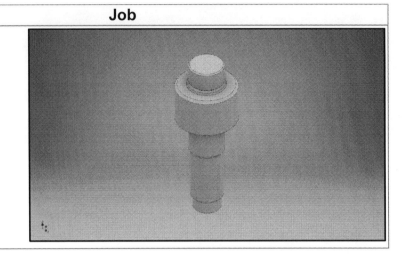

Total

NUMBER OF OPERATIONS: 7
NUMBER OF TOOLS: 3
TOOLS: T1 T2 T4
MAXIMUM Z: 8mm
MINIMUM Z: -277.3mm

MAXIMUM FEEDRATE: 1000mm/min
MAXIMUM SPINDLE SPEED: 500rpm
CUTTING DISTANCE: 7176.36mm
RAPID DISTANCE: 5373.2mm

Operation 1/7
DESCRIPTION: Face1
STRATEGY: Unspecified
WCS: #0
TOLERANCE: 0.01mm

MAXIMUM Z: 5mm
MINIMUM Z: 0mm
MAXIMUM SPINDLE SPEED: 500rpm
MAXIMUM FEEDRATE: 1000mm/min
CUTTING DISTANCE: 68.39mm
RAPID DISTANCE: 71.56mm
ESTIMATED CYCLE TIME: 5s (0.9%)
COOLANT: Flood

T1 D0 L0
TYPE: general turning
DIAMETER: 0mm
LENGTH: 0mm
FLUTES: 1

Operation 2/7
DESCRIPTION: Profile2
STRATEGY: Unspecified
WCS: #0
TOLERANCE: 0.01mm
STOCK TO LEAVE: 0mm
MAXIMUM STEPDOWN: 1mm
MAXIMUM STEPOVER: 1mm

MAXIMUM Z: 5mm
MINIMUM Z: -277.3mm
MAXIMUM SPINDLE SPEED: 500rpm
MAXIMUM FEEDRATE: 1000mm/min
CUTTING DISTANCE: 3244.27mm
RAPID DISTANCE: 3371.41mm
ESTIMATED CYCLE TIME: 3m:55s (42.4%)
COOLANT: Flood

T1 D0 L0
TYPE: general turning
DIAMETER: 0mm
LENGTH: 0mm
FLUTES: 1

Operation 3/7
DESCRIPTION: Groove2
STRATEGY: Unspecified
WCS: #0
TOLERANCE: 0.01mm

MAXIMUM Z: 5mm
MINIMUM Z: -255.47mm
MAXIMUM SPINDLE SPEED: 500rpm
MAXIMUM FEEDRATE: 1000mm/min

T2 D0 L0
TYPE: groove turning
DIAMETER: 0mm
LENGTH: 0mm

144

	Cutting Distance: 229.22mm	Flutes: 1
Stock to Leave: 0mm	Rapid Distance: 570.82mm	
Maximum stepover: 1mm	Estimated Cycle Time: 21s (3.7%)	
	Coolant: Flood	

Operation 4/7

Description: Groove4	Maximum Z: 5mm	**T2** D0 L0
Strategy: Unspecified	Minimum Z: -165.64mm	Type: groove turning
WCS: #0	Maximum Spindle Speed: 500rpm	Diameter: 0mm
Tolerance: 0.01mm	Maximum Feedrate: 1000mm/min	Length: 0mm
Stock to Leave: 0mm	Cutting Distance: 255.61mm	Flutes: 1
Maximum stepover: 1mm	Rapid Distance: 388.27mm	
	Estimated Cycle Time: 20s (3.6%)	
	Coolant: Flood	

Operation 5/7

Description: Groove6	Maximum Z: 8mm	**T4** D0 L0
Strategy: Unspecified	Minimum Z: -35mm	Type: groove turning
WCS: #0	Maximum Spindle Speed: 500rpm	Diameter: 0mm
Tolerance: 0.01mm	Maximum Feedrate: 1000mm/min	Length: 0mm
Stock to Leave: 0mm	Cutting Distance: 551.36mm	Flutes: 1
Maximum stepover: 1mm	Rapid Distance: 187.25mm	
	Estimated Cycle Time: 35s (6.4%)	
	Coolant: Flood	

Operation 6/7

Description: Groove7	Maximum Z: 8mm	**T4** D0 L0
Strategy: Unspecified	Minimum Z: -158.5mm	Type: groove turning
WCS: #0	Maximum Spindle Speed: 500rpm	Diameter: 0mm
Tolerance: 0.01mm	Maximum Feedrate: 1000mm/min	Length: 0mm
Stock to Leave: 0mm	Cutting Distance: 2738.47mm	Flutes: 1
Maximum stepover: 1mm	Rapid Distance: 689.67mm	
	Estimated Cycle Time: 2m:53s (31.1%)	
	Coolant: Flood	

Operation 7/7

Description: Face3	Maximum Z: 5mm	**T1** D0 L0
Strategy: Unspecified	Minimum Z: -1mm	Type: general turning
WCS: #0	Maximum Spindle Speed: 500rpm	Diameter: 0mm
Tolerance: 0.01mm	Maximum Feedrate: 1000mm/min	Length: 0mm
	Cutting Distance: 89.05mm	Flutes: 1
	Rapid Distance: 94.22mm	
	Estimated Cycle Time: 6s (1.2%)	
	Coolant: Flood	

Generated by Inventor HSM Pro 4.0.0.032

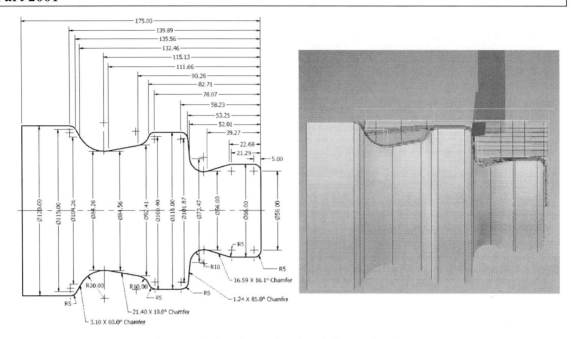

Figure 5.8 Drawing and tool path for part 2001

```
%
O2001 (Part 2001)
N10 G98 G18
N11 G21
N12 G50 S6000
N13 G28 U0.

(FACE5)
N14 T0101
N15 G54
N16 M8
N17 G98
N18 G97 S500 M3
N19 G0 X140. Z5.
N20 G0 Z-1.086
N21 G1 X122.828 F1000.
N22 X120. Z-2.5
N23 X-1.6
N24 X1.228 Z-1.086
N25 G0 X140.
N26 Z5.

(PROFILE1)
N27 G98
N28 G97 S500 M3
N29 G0 X140. Z5.
N30 Z1.404
N31 X121.821
N32 G1 X120.828 F1000.
N33 X118. Z-0.01
N34 Z-138.37
N35 G18 G3 X119.98 Z-
139.622 R6.8
N36 G1 X122.809 Z-
138.207
N37 G0 Z1.404
N38 X118.828
N39 G1 X116. Z-0.01
F1000.
```

```
N40 Z-137.548
N41 G3 X119. Z-138.927
R6.8
N42 G1 X121.828 Z-
137.513
N43 G0 Z1.404
N44 X116.828
N45 G1 X114. Z-0.01
F1000.
N46 Z-136.952
N47 X115.2 Z-137.298
N48 G3 X117. Z-137.919
R6.8
N49 G1 X119.828 Z-
136.505
N50 G0 Z1.404
N51 X114.828
N52 G1 X112. Z-0.01
F1000.
N53 Z-58.974
N54 G3 X113. Z-61.534
R6.8
N55 G1 Z-81.366
N56 G3 X112.948 Z-81.958
R6.8
N57 G1 X112. Z-87.378
N58 Z-136.374
N59 X115. Z-137.24
N60 X117.828 Z-135.826
N61 G0 Z1.404
N62 X112.828
N63 G1 X110. Z-0.01
F1000.
N64 Z-57.273
N65 G3 X113. Z-61.526
R6.8
N66 G1 X115.828 Z-60.112
N67 G0 Z1.404
N68 X110.828
```

```
N69 G1 X108. Z-0.01
F1000.
N70 Z-56.266
N71 G3 X111. Z-57.984
R6.8
N72 G1 X113.828 Z-56.57
N73 G0 Z1.404
N74 X108.828
N75 G1 X106. Z-0.01
F1000.
N76 Z-55.588
N77 G3 X109. Z-56.717
R6.8
N78 G1 X111.828 Z-55.303
N79 G0 Z1.404
N80 X106.828
N81 G1 X104. Z-0.01
F1000.
N82 Z-55.134
N83 G3 X107. Z-55.894
R6.8
N84 G1 X109.828 Z-54.48
N85 G0 Z1.404
N86 X104.828
N87 G1 X102. Z-0.01
F1000.
N88 Z-54.859
N89 G3 X105. Z-55.337
R6.8
N90 G1 X107.828 Z-53.922
N91 G0 Z1.404
N92 X102.828
N93 G1 X100. Z-0.01
F1000.
N94 Z-54.734
N95 X100.585 Z-54.759
N96 G3 X103. Z-54.976
R6.8
N97 G1 X105.828 Z-53.562
```

N98 G0 Z1.404
N99 X100.828
N100 G1 X98. Z-0.01
F1000.
N101 Z-54.646
N102 X100.585 Z-54.759
N103 G3 X101. Z-54.781
R6.8
N104 G1 X103.828 Z-
53.366
N105 G0 Z1.404
N106 X98.828
N107 G1 X96. Z-0.01
F1000.
N108 Z-54.559
N109 X99. Z-54.69
N110 X101.828 Z-53.276
N111 G0 Z1.404
N112 X96.828
N113 G1 X94. Z-0.01
F1000.
N114 Z-54.471
N115 X97. Z-54.603
N116 X99.828 Z-53.188
N117 G0 Z1.404
N118 X94.828
N119 G1 X92. Z-0.01
F1000.
N120 Z-54.384
N121 X95. Z-54.515
N122 X97.828 Z-53.101
N123 G0 Z1.404
N124 X92.828
N125 G1 X90. Z-0.01
F1000.
N126 Z-54.296
N127 X93. Z-54.428
N128 X95.828 Z-53.013
N129 G0 Z1.404
N130 X90.828
N131 G1 X88. Z-0.01
F1000.
N132 Z-54.209
N133 X91. Z-54.34
N134 X93.828 Z-52.926
N135 G0 Z1.404
N136 X88.828
N137 G1 X86. Z-0.01
F1000.
N138 Z-54.121
N139 X89. Z-54.253
N140 X91.828 Z-52.838
N141 G0 Z1.404
N142 X86.828
N143 G1 X84. Z-0.01
F1000.
N144 Z-54.034
N145 X87. Z-54.165
N146 X89.828 Z-52.751
N147 G0 Z1.404
N148 X84.828
N149 G1 X82. Z-0.01
F1000.
N150 Z-53.946
N151 X85. Z-54.078
N152 X87.828 Z-52.663
N153 G0 Z1.404
N154 X82.828
N155 G1 X80. Z-0.01
F1000.
N156 Z-53.859
N157 X83. Z-53.99
N158 X85.828 Z-52.576

N159 G0 Z1.404
N160 X80.828
N161 G1 X78. Z-0.01
F1000.
N162 Z-53.771
N163 X81. Z-53.903
N164 X83.828 Z-52.488
N165 G0 Z1.404
N166 X78.828
N167 G1 X76. Z-0.01
F1000.
N168 Z-53.684
N169 X79. Z-53.815
N170 X81.828 Z-52.401
N171 G0 Z1.404
N172 X76.828
N173 G1 X74. Z-0.01
F1000.
N174 Z-53.596
N175 X77. Z-53.728
N176 X79.828 Z-52.313
N177 G0 Z1.404
N178 X74.828
N179 G1 X72. Z-0.01
F1000.
N180 Z-53.508
N181 G2 X72.181 Z-53.517
R8.2
N182 G1 X75. Z-53.64
N183 X77.828 Z-52.226
N184 G0 Z1.404
N185 X72.828
N186 G1 X70. Z-0.01
F1000.
N187 Z-53.347
N188 G2 X72.181 Z-53.517
R8.2
N189 G1 X73. Z-53.553
N190 X75.828 Z-52.138
N191 G0 Z1.404
N192 X70.828
N193 G1 X68. Z-0.01
F1000.
N194 Z-8.281
N195 G3 Z-8.3 R6.8
N196 G1 Z-24.591
N197 G3 Z-24.609 R6.8
N198 G1 Z-53.053
N199 G2 X71. Z-53.443
R8.2
N200 G1 X73.828 Z-52.029
N201 G0 Z1.404
N202 X68.828
N203 G1 X66. Z-0.01
F1000.
N204 Z-4.75
N205 G3 X68. Z-8.3 R6.8
N206 G1 X70.828 Z-6.886
N207 G0 Z1.404
N208 X66.828
N209 G1 X64. Z-0.01
F1000.
N210 Z-3.483
N211 G3 X67. Z-5.741
R6.8
N212 G1 X69.828 Z-4.326
N213 G0 Z1.404
N214 X64.828
N215 G1 X62. Z-0.01
F1000.
N216 Z-2.661
N217 G3 X65. Z-4.04 R6.8
N218 G1 X67.828 Z-2.625

N219 G0 Z1.404
N220 X62.828
N221 G1 X60. Z-0.01
F1000.
N222 Z-2.103
N223 G3 X63. Z-3.032
R6.8
N224 G1 X65.828 Z-1.618
N225 G0 Z1.404
N226 X60.828
N227 G1 X58. Z-0.01
F1000.
N228 Z-1.743
N229 G3 X61. Z-2.354
R6.8
N230 G1 X63.828 Z-0.94
N231 G0 Z1.404
N232 X58.828
N233 G1 X56. Z-0.01
F1000.
N234 Z-1.547
N235 G3 X59. Z-1.901
R6.8
N236 G1 X61.828 Z-0.487
N237 G0 Z1.404
N238 X56.828
N239 G1 X54. Z-0.01
F1000.
N240 Z-1.5
N241 X54.4
N242 G3 X57. Z-1.625
R6.8
N243 G1 X59.828 Z-0.211
N244 G0 Z1.404
N245 X54.828
N246 G1 X52. Z-0.01
F1000.
N247 Z-1.5
N248 X54.4
N249 G3 X55. Z-1.507
R6.8
N250 G1 X57.828 Z-0.092
N251 G0 Z1.404
N252 X52.828
N253 G1 X50. Z-0.01
F1000.
N254 Z-1.5
N255 X53.
N256 X55.828 Z-0.086
N257 G0 Z1.404
N258 X50.828
N259 G1 X48. Z-0.01
F1000.
N260 Z-1.5
N261 X51.
N262 X53.828 Z-0.086
N263 G0 Z1.404
N264 X48.828
N265 G1 X46. Z-0.01
F1000.
N266 Z-1.5
N267 X49.
N268 X51.828 Z-0.086
N269 G0 Z1.404
N270 X46.828
N271 G1 X44. Z-0.01
F1000.
N272 Z-1.5
N273 X47.
N274 X49.828 Z-0.086
N275 G0 Z1.404
N276 X44.828
N277 G1 X42. Z-0.01

```
F1000.
N278 Z-1.5
N279 X45.
N280 X47.828 Z-0.086
N281 G0 Z1.404
N282 X42.828
N283 G1 X40. Z-0.01
F1000.
N284 Z-1.5
N285 X43.
N286 X45.828 Z-0.086
N287 G0 Z1.404
N288 X40.828
N289 G1 X38. Z-0.01
F1000.
N290 Z-1.5
N291 X41.
N292 X43.828 Z-0.086
N293 G0 Z1.404
N294 X38.828
N295 G1 X36. Z-0.01
F1000.
N296 Z-1.5
N297 X39.
N298 X41.828 Z-0.086
N299 G0 Z1.404
N300 X36.828
N301 G1 X34. Z-0.01
F1000.
N302 Z-1.5
N303 X37.
N304 X39.828 Z-0.086
N305 G0 Z1.404
N306 X34.828
N307 G1 X32. Z-0.01
F1000.
N308 Z-1.5
N309 X35.
N310 X37.828 Z-0.086
N311 G0 Z1.404
N312 X32.828
N313 G1 X30. Z-0.01
F1000.
N314 Z-1.5
N315 X33.
N316 X35.828 Z-0.086
N317 G0 Z1.404
N318 X30.828
N319 G1 X28. Z-0.01
F1000.
N320 Z-1.5
N321 X31.
N322 X33.828 Z-0.086
N323 G0 Z1.404
N324 X28.828
N325 G1 X26. Z-0.01
F1000.
N326 Z-1.5
N327 X29.
N328 X31.828 Z-0.086
N329 G0 Z1.404
N330 X26.828
N331 G1 X24. Z-0.01
F1000.
N332 Z-1.5
N333 X27.
N334 X29.828 Z-0.086
N335 G0 Z1.404
N336 X24.828
N337 G1 X22. Z-0.01
F1000.
N338 Z-1.5
N339 X25.

N340 X27.828 Z-0.086
N341 G0 Z1.404
N342 X22.828
N343 G1 X20. Z-0.01
F1000.
N344 Z-1.5
N345 X23.
N346 X25.828 Z-0.086
N347 G0 Z1.404
N348 X20.828
N349 G1 X18. Z-0.01
F1000.
N350 Z-1.5
N351 X21.
N352 X23.828 Z-0.086
N353 G0 Z1.404
N354 X18.828
N355 G1 X16. Z-0.01
F1000.
N356 Z-1.5
N357 X19.
N358 X21.828 Z-0.086
N359 G0 Z1.404
N360 X16.828
N361 G1 X14. Z-0.01
F1000.
N362 Z-1.5
N363 X17.
N364 X19.828 Z-0.086
N365 G0 Z1.404
N366 X14.828
N367 G1 X12. Z-0.01
F1000.
N368 Z-1.5
N369 X15.
N370 X17.828 Z-0.086
N371 G0 Z1.404
N372 X12.828
N373 G1 X10. Z-0.01
F1000.
N374 Z-1.5
N375 X13.
N376 X15.828 Z-0.086
N377 G0 Z1.404
N378 X10.828
N379 G1 X8. Z-0.01
F1000.
N380 Z-1.5
N381 X11.
N382 X13.828 Z-0.086
N383 G0 Z1.404
N384 X8.828
N385 G1 X6. Z-0.01
F1000.
N386 Z-1.5
N387 X9.
N388 X11.828 Z-0.086
N389 G0 Z1.404
N390 X6.828
N391 G1 X4. Z-0.01
F1000.
N392 Z-1.5
N393 X7.
N394 X9.828 Z-0.086
N395 G0 Z1.404
N396 X4.828
N397 G1 X2. Z-0.01
F1000.
N398 Z-1.5
N399 X5.
N400 X7.828 Z-0.086
N401 G0 Z1.404
N402 X2.828

N403 G1 X0. Z-0.01
F1000.
N404 Z-1.5
N405 X3.
N406 X5.828 Z-0.086
N407 X60.709
N408 G0 X70.828
N409 Z-23.195
N410 G1 X68. Z-24.609
F1000.
N411 G3 X67.948 Z-25.183
R6.8
N412 G1 X66. Z-36.318
N413 Z-52.612
N414 G2 X69. Z-53.217
R8.2
N415 G1 X71.828 Z-51.803
N416 G0 Z-34.904
N417 X70.263
N418 G1 X68.828 F1000.
N419 X66. Z-36.318
N420 X64.652 Z-44.022
N421 Z-52.216
N422 G2 X67. Z-52.852
R8.2
N423 G1 X69.828 Z-51.438
N424 G0 Z-42.608
N425 X68.915
N426 G1 X67,48 F1000.
N427 X64.652 Z-44.022
N428 X63.304 Z-51.726
N429 G2 X65.652 Z-52.518
R8.2
N430 G1 X68.48 Z-51.104
N431 X69.003
N432 G0 X114.828
N433 Z-85.963
N434 G1 X112. Z-87.378
F1000.
N435 X110. Z-98.808
N436 Z-135.797
N437 X113. Z-136.663
N438 X115.828 Z-135.249
N439 G0 Z-97.394
N440 X114.263
N441 G1 X112.828 F1000.
N442 X110. Z-98.808
N443 X108. Z-110.238
N444 Z-135.219
N445 X111. Z-136.086
N446 X113.828 Z-134.671
N447 G0 Z-108.824
N448 X112.263
N449 G1 X110.828 F1000.
N450 X108. Z-110.238
N451 X106. Z-121.668
N452 Z-134.642
N453 X109. Z-135.508
N454 X111.828 Z-134.094
N455 G0 Z-120.254
N456 X110.263
N457 G1 X108.828 F1000.
N458 X106. Z-121.668
N459 X104.92 Z-127.841
N460 Z-134.33
N461 X107. Z-134.931
N462 X109.828 Z-133.516
N463 G0 Z-126.427
N464 X109.183
N465 G1 X107.748 F1000.
N466 X104.92 Z-127.841
N467 X103.84 Z-134.014
N468 G2 X104.456 Z-
```

134.196 R18.2
N469 G1 X105.92 Z-
134.619
N470 X108.748 Z-133.205
N471 G0 X113.2
N472 Z-1.086
N473 X60.326
N474 G1 X1.228 F1000.
N475 X-1.6 Z-2.5
N476 X54.4
N477 G3 X66. Z-8.3 R5.8
N478 G1 Z-24.591
N479 G3 X65.956 Z-25.096
R5.8
N480 G1 X61.222 Z-52.15
N481 G2 X72.007 Z-54.513
R9.2
N482 G1 X100.411 Z-
55.756
N483 G3 X111. Z-61.534
R5.8
N484 G1 Z-81.366
N485 G3 X110.956 Z-
81.871 R5.8
N486 G1 X101.741 Z-
134.537
N487 G2 X103.456 Z-
135.062 R19.2
N488 G1 X114.2 Z-138.164
N489 G3 X119.982 Z-
142.863 R5.8
N490 G1 X122.81 Z-
141.449
N491 X123.607
N492 G0 X140.
N493 Z5.
N494 G28 U0.

(GROOVE5)
N495 M1
N496 T0202
N497 G54
N498 G98
N499 G97 S500 M3
N500 G0 X140. Z5.
N501 G0 Z-12.5
N502 X124.
N503 G1 X120. F1000.
N504 X68.
N505 X124.
N506 G0 Z-13.477
N507 G1 X120. F1000.
N508 X68.
N509 X123.49
N510 G0 X124.
N511 Z-14.454
N512 G1 X120. F1000.
N513 X68.
N514 X118.578
N515 X120.
N516 X123.489
N517 G0 X124.
N518 Z-15.431
N519 G1 X120. F1000.
N520 X68.
N521 X70. Z-14.431
N522 X121.029
N523 G0 X124.
N524 Z-16.408
N525 G1 X120. F1000.
N526 X68.
N527 X70. Z-15.408
N528 X120.593

N529 G0 X124.
N530 Z-17.385
N531 G1 X120. F1000.
N532 X68.
N533 X70. Z-16.385
N534 X120.593
N535 G0 X124.
N536 Z-18.362
N537 G1 X120. F1000.
N538 X68.
N539 X70. Z-17.362
N540 X120.593
N541 G0 X124.
N542 Z-19.339
N543 G1 X120. F1000.
N544 X68.
N545 X70. Z-18.339
N546 X120.593
N547 G0 X124.
N548 Z-20.316
N549 G1 X120. F1000.
N550 X68.
N551 X70. Z-19.316
N552 X120.593
N553 G0 X124.
N554 Z-21.293
N555 G1 X120. F1000.
N556 X68.
N557 X70. Z-20.293
N558 X120.593
N559 G0 X124.
N560 Z-22.27
N561 G1 X120. F1000.
N562 X68.
N563 X70. Z-21.27
N564 X120.593
N565 G0 X124.
N566 Z-23.247
N567 G1 X120. F1000.
N568 X68.
N569 X70. Z-22.247
N570 X120.593
N571 G0 X124.
N572 Z-24.224
N573 G1 X120. F1000.
N574 X68.
N575 X70. Z-23.224
N576 X120.593
N577 G0 X124.
N578 Z-25.201
N579 G1 X120. F1000.
N580 X68.
N581 X70. Z-24.201
N582 X120.593
N583 G0 X124.
N584 Z-26.178
N585 G1 X120. F1000.
N586 X68.
N587 X70. Z-25.178
N588 X120.593
N589 G0 X124.
N590 Z-27.155
N591 G1 X120. F1000.
N592 X68.
N593 X70. Z-26.155
N594 X120.593
N595 G0 X124.
N596 Z-28.132
N597 G1 X120. F1000.
N598 X68.
N599 X70. Z-27.132
N600 X120.58
N601 G0 X124.

N602 Z-29.109
N603 G1 X120. F1000.
N604 X67.991
N605 X68.194
N606 X70.194 Z-28.109
N607 X120.58
N608 G0 X124.
N609 Z-30.086
N610 G1 X120. F1000.
N611 X67.847
N612 X68.051
N613 X70.051 Z-29.086
N614 X120.58
N615 G0 X124.
N616 Z-31.063
N617 G1 X120. F1000.
N618 X67.526
N619 X68.141
N620 X70.141 Z-30.063
N621 X120.58
N622 G0 X124.
N623 Z-32.04
N624 G1 X120. F1000.
N625 X67.022
N626 X67.85
N627 X69.85 Z-31.04
N628 X120.58
N629 G0 X124.
N630 Z-33.017
N631 G1 X120. F1000.
N632 X66.456
N633 X67.711
N634 X69.711 Z-32.017
N635 X120.58
N636 G0 X124.
N637 Z-33.993
N638 G1 X120. F1000.
N639 X65.891
N640 X67.159
N641 X69.159 Z-32.993
N642 X120.58
N643 G0 X124.
N644 Z-34.97
N645 G1 X120. F1000.
N646 X65.325
N647 X66.606
N648 X68.606 Z-33.97
N649 X120.58
N650 G0 X124.
N651 Z-35.947
N652 G1 X120. F1000.
N653 X64.759
N654 X66.054
N655 X68.054 Z-34.947
N656 X120.58
N657 G0 X124.
N658 Z-36.924
N659 G1 X120. F1000.
N660 X64.193
N661 X65.501
N662 X67.501 Z-35.924
N663 X120.58
N664 G0 X124.
N665 Z-37.901
N666 G1 X120. F1000.
N667 X63.627
N668 X64.949
N669 X66.949 Z-36.901
N670 X120.58
N671 G0 X124.
N672 Z-38.878
N673 G1 X120. F1000.
N674 X63.062

```
N675 X64.174
N676 X66.174 Z-37.878
N677 X120.58
N678 G0 X124.
N679 Z-39.855
N680 G1 X120. F1000.
N681 X62.496
N682 X63.619
N683 X65.619 Z-38.855
N684 X120.58
N685 G0 X124.
N686 Z-40.832
N687 G1 X120. F1000.
N688 X61.93
N689 X63.064
N690 X65.064 Z-39.832
N691 X120.58
N692 G0 X124.
N693 Z-41.809
N694 G1 X120. F1000.
N695 X61.364
N696 X62.509
N697 X64.509 Z-40.809
N698 X120.58
N699 G0 X124.
N700 Z-42.786
N701 G1 X120. F1000.
N702 X60.798
N703 X61.955
N704 X63.955 Z-41.786
N705 X120.58
N706 G0 X124.
N707 Z-43.763
N708 G1 X120. F1000.
N709 X60.232
N710 X61.4
N711 X63.4 Z-42.763
N712 X120.58
N713 G0 X124.
N714 Z-44.74
N715 G1 X120. F1000.
N716 X59.667
N717 X60.845
N718 X62.845 Z-43.74
N719 X120.58
N720 G0 X124.
N721 Z-45.717
N722 G1 X120. F1000.
N723 X59.101
N724 X60.29
N725 X62.29 Z-44.717
N726 X120.58
N727 G0 X124.
N728 Z-46.694
N729 G1 X120. F1000.
N730 X58.535
N731 X59.735
N732 X61.735 Z-45.694
N733 X120.58
N734 G0 X124.
N735 Z-47.671
N736 G1 X120. F1000.
N737 X57.969
N738 X59.181
N739 X61.181 Z-46.671
N740 X120.58
N741 G0 X124.
N742 Z-48.648
N743 G1 X120. F1000.
N744 X57.416
N745 X58.638
N746 X60.638 Z-47.648
N747 X120.58

N748 G0 X124.
N749 Z-49.625
N750 G1 X120. F1000.
N751 X57.212
N752 X58.193
N753 X60.193 Z-48.625
N754 X120.593
N755 G0 X124.
N756 Z-50.602
N757 G1 X120. F1000.
N758 X57.494
N759 X59.494 Z-49.602
N760 X120.593
N761 G0 X124.
N762 Z-51.579
N763 G1 X120. F1000.
N764 X58.319
N765 X60.319 Z-50.579
N766 X120.593
N767 G0 X124.
N768 Z-52.556
N769 G1 X120. F1000.
N770 X59.937
N771 X61.937 Z-51.556
N772 X120.593
N773 G0 X124.
N774 Z-53.533
N775 G1 X120. F1000.
N776 X64.518
N777 X66.518 Z-52.533
N778 X90.314
N779 G0 X124.
N780 Z-54.51
N781 G1 X120. F1000.
N782 X86.852
N783 X88.852 Z-53.51
N784 G0 Z-12.5
N785 G1 X72. F1000.
N786 X68.
N787 Z-28.791
N788 G18 G3 X67.132 Z-
31.85 R11.
N789 G1 X57.526 Z-48.435
N790 G2 X57.211 Z-49.548
R4.
N791 G1 X60.039 Z-48.134
N792 X61.865
N793 G0 X90.854
N794 Z-54.51
N795 G1 X90.852 F1000.
N796 X86.852
N797 X64.513 Z-53.533
N798 G3 X57.211 Z-49.548
R4.
N799 G1 X60.039 Z-50.962
N800 X62.385
N801 G0 X72.002
N802 Z-12.5
N803 G1 X70. F1000.
N804 X66.
N805 Z-28.791
N806 G3 X65.211 Z-31.572
R10.
N807 G1 X55.605 Z-48.157
N808 G2 X63.979 Z-54.51
R5.
N809 G1 X66.808 Z-53.096
N810 G0 X140.
N811 Z5.
N812 G28 U0.

(GROOVE6)
N813 M1

N814 T0303
N815 G54
N816 G98
N817 G97 S500 M3
N818 G0 X140. Z5.
N819 G0 Z-83.566
N820 X124.
N821 G1 X120. F1000.
N822 X112.697
N823 X124.
N824 G0 Z-84.554
N825 G1 X120. F1000.
N826 X112.15
N827 X123.478
N828 G0 X124.
N829 Z-85.542
N830 G1 X120. F1000.
N831 X111.291
N832 X118.333
N833 X120.
N834 X123.475
N835 G0 X124.
N836 Z-86.53
N837 G1 X120. F1000.
N838 X110.053
N839 X112.151
N840 X114.151 Z-85.53
N841 X120.737
N842 G0 X124.
N843 Z-87.519
N844 G1 X120. F1000.
N845 X108.299
N846 X111.315
N847 X113.315 Z-86.519
N848 X120.395
N849 G0 X124.
N850 Z-88.507
N851 G1 X120. F1000.
N852 X105.684
N853 X110.046
N854 X112.046 Z-87.507
N855 X120.395
N856 G0 X124.
N857 Z-89.495
N858 G1 X120. F1000.
N859 X101.639
N860 X108.381
N861 X110.381 Z-88.495
N862 X120.395
N863 G0 X124.
N864 Z-90.483
N865 G1 X120. F1000.
N866 X98.944
N867 X106.675
N868 X108.675 Z-89.483
N869 X120.395
N870 G0 X124.
N871 Z-91.472
N872 G1 X120. F1000.
N873 X97.149
N874 X104.022
N875 X106.022 Z-90.472
N876 X120.395
N877 G0 X124.
N878 Z-92.46
N879 G1 X120. F1000.
N880 X95.882
N881 X100.027
N882 X102.027 Z-91.46
N883 X120.395
N884 G0 X124.
N885 Z-93.448
N886 G1 X120. F1000.
```

```
N887 X95.001
N888 X97.344
N889 X99.344 Z-92.448
N890 X120.395
N891 G0 X124.
N892 Z-94.437
N893 G1 X120. F1000.
N894 X94.436
N895 X95.933
N896 X97.933 Z-93.437
N897 X120.395
N898 G0 X124.
N899 Z-95.425
N900 G1 X120. F1000.
N901 X94.077
N902 X94.989
N903 X96.989 Z-94.425
N904 X120.395
N905 G0 X124.
N906 Z-96.413
N907 G1 X120. F1000.
N908 X93.729
N909 X94.447
N910 X96.447 Z-95.413
N911 X120.395
N912 G0 X124.
N913 Z-97.401
N914 G1 X120. F1000.
N915 X93.38
N916 X94.108
N917 X96.108 Z-96.401
N918 X120.395
N919 G0 X124.
N920 Z-98.39
N921 G1 X120. F1000.
N922 X93.032
N923 X93.769
N924 X95.769 Z-97.39
N925 X120.395
N926 G0 X124.
N927 Z-99.378
N928 G1 X120. F1000.
N929 X92.683
N930 X93.43
N931 X95.43 Z-98.378
N932 X120.395
N933 G0 X124.
N934 Z-100.366
N935 G1 X120. F1000.
N936 X92.335
N937 X93.091
N938 X95.091 Z-99.366
N939 X120.395
N940 G0 X124.
N941 Z-101.354
N942 G1 X120. F1000.
N943 X91.986
N944 X92.752
N945 X94.752 Z-100.354
N946 X120.395
N947 G0 X124.
N948 Z-102.343
N949 G1 X120. F1000.
N950 X91.638
N951 X92.413
N952 X94.413 Z-101.343
N953 X120.395
N954 G0 X124.
N955 Z-103.331
N956 G1 X120. F1000.
N957 X91.289
N958 X92.074
N959 X94.074 Z-102.331

N960 X120.395
N961 G0 X124.
N962 Z-104.319
N963 G1 X120. F1000.
N964 X90.941
N965 X91.622
N966 X93.622 Z-103.319
N967 X120.395
N968 G0 X124.
N969 Z-105.307
N970 G1 X120. F1000.
N971 X90.592
N972 X91.282
N973 X93.282 Z-104.307
N974 X120.395
N975 G0 X124.
N976 Z-106.296
N977 G1 X120. F1000.
N978 X90.244
N979 X90.941
N980 X92.941 Z-105.296
N981 X120.395
N982 G0 X124.
N983 Z-107.284
N984 G1 X120. F1000.
N985 X89.895
N986 X90.601
N987 X92.601 Z-106.284
N988 X120.395
N989 G0 X124.
N990 Z-108.272
N991 G1 X120. F1000.
N992 X89.547
N993 X90.261
N994 X92.261 Z-107.272
N995 X120.395
N996 G0 X124.
N997 Z-109.26
N998 G1 X120. F1000.
N999 X89.198
N1000 X89.92
N1001 X91.92 Z-108.26
N1002 X120.395
N1003 G0 X124.
N1004 Z-110.249
N1005 G1 X120. F1000.
N1006 X88.85
N1007 X89.58
N1008 X91.58 Z-109.249
N1009 X120.395
N1010 G0 X124.
N1011 Z-111.237
N1012 G1 X120. F1000.
N1013 X88.501
N1014 X89.24
N1015 X91.24 Z-110.237
N1016 X120.395
N1017 G0 X124.
N1018 Z-112.225
N1019 G1 X120. F1000.
N1020 X88.153
N1021 X88.899
N1022 X90.899 Z-111.225
N1023 X120.395
N1024 G0 X124.
N1025 Z-113.213
N1026 G1 X120. F1000.
N1027 X87.804
N1028 X88.559
N1029 X90.559 Z-112.213
N1030 X120.395
N1031 G0 X124.
N1032 Z-114.202

N1033 G1 X120. F1000.
N1034 X87.456
N1035 X88.218
N1036 X90.218 Z-113.202
N1037 X120.395
N1038 G0 X124.
N1039 Z-115.19
N1040 G1 X120. F1000.
N1041 X87.107
N1042 X87.878
N1043 X89.878 Z-114.19
N1044 X120.395
N1045 G0 X124.
N1046 Z-116.178
N1047 G1 X120. F1000.
N1048 X86.759
N1049 X87.538
N1050 X89.538 Z-115.178
N1051 X120.395
N1052 G0 X124.
N1053 Z-117.166
N1054 G1 X120. F1000.
N1055 X86.478
N1056 X87.133
N1057 X89.133 Z-116.166
N1058 X120.395
N1059 G0 X124.
N1060 Z-118.155
N1061 G1 X120. F1000.
N1062 X86.311
N1063 X86.837
N1064 X88.837 Z-117.155
N1065 X120.395
N1066 G0 X124.
N1067 Z-119.143
N1068 G1 X120. F1000.
N1069 X86.256
N1070 X86.52
N1071 X88.52 Z-118.143
N1072 X120.416
N1073 G0 X124.
N1074 Z-120.131
N1075 G1 X120. F1000.
N1076 X86.313
N1077 X88.313 Z-119.131
N1078 X120.416
N1079 G0 X124.
N1080 Z-121.119
N1081 G1 X120. F1000.
N1082 X86.482
N1083 X88.482 Z-120.119
N1084 X120.416
N1085 G0 X124.
N1086 Z-122.108
N1087 G1 X120. F1000.
N1088 X86.765
N1089 X88.765 Z-121.108
N1090 X120.416
N1091 G0 X124.
N1092 Z-123.096
N1093 G1 X120. F1000.
N1094 X87.165
N1095 X89.165 Z-122.096
N1096 X120.416
N1097 G0 X124.
N1098 Z-124.084
N1099 G1 X120. F1000.
N1100 X87.685
N1101 X89.685 Z-123.084
N1102 X120.416
N1103 G0 X124.
N1104 Z-125.072
N1105 G1 X120. F1000.
```

```
N1106 X88.332                N1142 X95.533                 N1175 G2 X86.256 Z-
N1107 X90.332 Z-124.072      N1143 X97.533 Z-130.002        119.135 R17.5
N1108 X120.416               N1144 X120.416                N1176 G1 X89.084 Z-
N1109 G0 X124.               N1145 G0 X124.                 117.72
N1110 Z-126.061              N1146 Z-131.99                N1177 X90.387
N1111 G1 X120. F1000.        N1147 G1 X120. F1000.         N1178 G0 X110.061
N1112 X89.114                N1148 X97.509                 N1179 Z-134.955
N1113 X91.114 Z-125.061      N1149 X99.509 Z-130.99        N1180 G1 X110.059 F1000.
N1114 X120.416               N1150 X120.416                N1181 X106.059
N1115 G0 X124.               N1151 G0 X124.                N1182 X103.756 Z-134.29
N1116 Z-127.049              N1152 Z-132.978               N1183 G3 X86.256 Z-
N1117 G1 X120. F1000.        N1153 G1 X120. F1000.          119.135 R17.5
N1118 X90.04                 N1154 X99.846                 N1184 G1 X89.084 Z-
N1119 X92.04 Z-126.049       N1155 X101.846 Z-131.978       120.549
N1120 X120.416               N1156 X120.416                N1185 X90.386
N1121 G0 X124.               N1157 G0 X124.                N1186 G0 X116.699
N1122 Z-128.037              N1158 Z-133.967               N1187 Z-83.566
N1123 G1 X120. F1000.        N1159 G1 X120. F1000.         N1188 G1 X114.649 F1000.
N1124 X91.123                N1160 X102.681                N1189 X110.649
N1125 X93.123 Z-127.037      N1161 X104.681 Z-132.967      N1190 G3 X102.813 Z-
N1126 X120.416               N1162 X109.521                 88.104 R6.5
N1127 G0 X124.               N1163 G0 X124.                N1191 G2 X92.364 Z-
N1128 Z-129.025              N1164 Z-134.955                94.524 R8.5
N1129 G1 X120. F1000.        N1165 G1 X120. F1000.         N1192 G1 X84.818 Z-
N1130 X92.383                N1166 X106.059                 115.922
N1131 X94.383 Z-128.025      N1167 X108.059 Z-133.955      N1193 G2 X102.077 Z-
N1132 X120.416               N1168 X116.699                 134.955 R18.5
N1133 G0 X124.               N1169 G0 Z-83.566             N1194 G1 X104.905 Z-
N1134 Z-130.014              N1170 G1 X116.697 F1000.       133.541
N1135 G1 X120. F1000.        N1171 X112.697                N1195 G0 X140.
N1136 X93.841                N1172 G18 G3 X103.553 Z-      N1196 Z5.
N1137 X95.841 Z-129.014       89.033 R7.5
N1138 X120.416               N1173 G2 X94.334 Z-           N1197 M9
N1139 G0 X124.                94.698 R7.5                  N1198 G28 U0. W0.
N1140 Z-131.002              N1174 G1 X86.788 Z-           N1199 M30
N1141 G1 X120. F1000.         116.096                      %
```

Setup Sheet - -Part 001

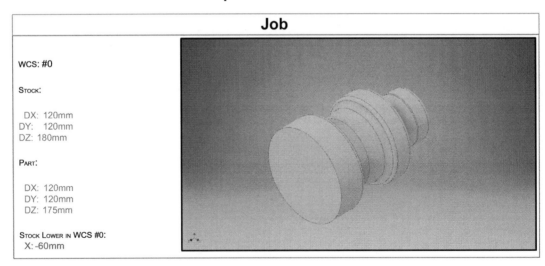

Job
WCS: #0
STOCK:
DX: 120mm
DY: 120mm
DZ: 180mm
PART:
DX: 120mm
DY: 120mm
DZ: 175mm
STOCK LOWER IN WCS #0:
X: -60mm

Total

NUMBER OF OPERATIONS: 4
NUMBER OF TOOLS: 3
TOOLS: **T1 T2 T3**
MAXIMUM Z: 5mm
MINIMUM Z: -142.86mm

MAXIMUM FEEDRATE: 1000mm/min
MAXIMUM SPINDLE SPEED: 500rpm
CUTTING DISTANCE: 6991.55mm
RAPID DISTANCE: 3549.51mm

Operation 1/4

DESCRIPTION: Face5	MAXIMUM Z: 5mm	**T1** D0 L0
STRATEGY: Unspecified	MINIMUM Z: -2.5mm	TYPE: general turning
WCS: #0	MAXIMUM SPINDLE SPEED: 500rpm	DIAMETER: 0mm
TOLERANCE: 0.01mm	MAXIMUM FEEDRATE: 1000mm/min	LENGTH: 0mm
	CUTTING DISTANCE: 73.39mm	FLUTES: 1
	RAPID DISTANCE: 81.56mm	
	ESTIMATED CYCLE TIME: 5s (1.1%)	
	COOLANT: Flood	

Operation 2/4

DESCRIPTION: Profile1	MAXIMUM Z: 5mm	**T1** D0 L0
STRATEGY: Unspecified	MINIMUM Z: -142.86mm	TYPE: general turning
WCS: #0	MAXIMUM SPINDLE SPEED: 500rpm	DIAMETER: 0mm
TOLERANCE: 0.01mm	MAXIMUM FEEDRATE: 1000mm/min	LENGTH: 0mm
STOCK TO LEAVE: 0mm	CUTTING DISTANCE: 2631.14mm	FLUTES: 1
MAXIMUM STEPDOWN: 1mm	RAPID DISTANCE: 2502.29mm	
MAXIMUM STEPOVER: 1mm	ESTIMATED CYCLE TIME: 3m:8s (37.1%)	
	COOLANT: Flood	

Operation 3/4

DESCRIPTION: Groove5	MAXIMUM Z: 5mm	**T2** D0 L0
	MINIMUM Z: -54.51mm	TYPE: groove turning
STRATEGY: Unspecified		DIAMETER: 0mm
	MAXIMUM SPINDLE SPEED: 500rpm	LENGTH: 0mm
WCS: #0	MAXIMUM FEEDRATE: 1000mm/min	
	CUTTING DISTANCE: 2631.63mm	
	RAPID DISTANCE: 391.07mm	

Operation 4/4

DESCRIPTION: Groove6	MAXIMUM Z: 5mm	**T3** D0 L0
STRATEGY: Unspecified	MINIMUM Z: -134.96mm	TYPE: groove turning
WCS: #0	MAXIMUM SPINDLE SPEED: 500rpm	DIAMETER: 0mm
TOLERANCE: 0.01mm	MAXIMUM FEEDRATE: 1000mm/min	LENGTH: 0mm
STOCK TO LEAVE: 0mm	CUTTING DISTANCE: 1655.4mm	FLUTES: 1

Generated by Inventor HSM Pro 4.0.0.032

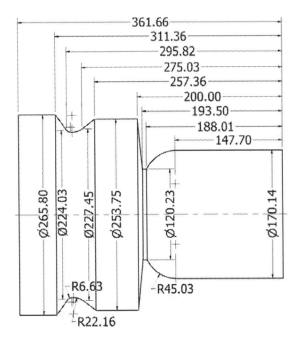
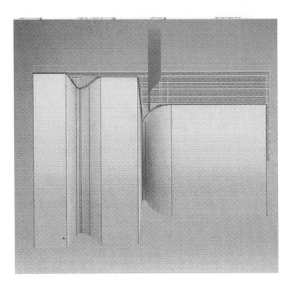

Figure 5.9 Drawing and tool path for part 9004

```
%
O9004 (Part 9004)
N10 G98 G18
N11 G21
N12 G50 S6000
N13 M31
N14 G53 G0 X0.

(Profile1)
N15 T101
N16 G98
N17 M22
N18 G97 S500 M3
N19 G54
N20 M8
N21 G0 X290. Z5.
N22 G0 Z1.404
N23 X271.821
N24 G1 X268.828 F1000.
N25 X266. Z-0.01
N26 Z-314.759
N27 X266.349 Z-314.889
N28 G18 G3 X267.8 Z-
316.333 R1.8
N29 G1 Z-366.631
N30 X269.98
N31 X272.808 Z-365.217
N32 G0 Z1.404
N33 X264.828
N34 G1 X262. Z-0.01 F1000.
N35 Z-313.271
N36 X266.349 Z-314.889
N37 G3 X267.8 Z-316.333
R1.8
N38 G1 X270.628 Z-314.919
N39 G0 Z1.404
```

```
N40 X260.828
N41 G1 X258. Z-0.01 F1000.
N42 Z-311.782
N43 X264. Z-314.015
N44 X266.828 Z-312.601
N45 G0 Z1.404
N46 X256.828
N47 G1 X254. Z-0.01 F1000.
N48 Z-203.426
N49 G3 X255.746 Z-204.969
R1.8
N50 G1 Z-262.329
N51 G3 X255.732 Z-262.486
R1.8
N52 G1 X254. Z-272.386
N53 Z-310.294
N54 X260. Z-312.526
N55 X262.828 Z-311.112
N56 G0 Z1.404
N57 X252.828
N58 G1 X250. Z-0.01 F1000.
N59 Z-203.056
N60 X252.495 Z-203.177
N61 G3 X255.746 Z-204.969
R1.8
N62 G1 X258.574 Z-203.555
N63 G0 Z1.404
N64 X248.828
N65 G1 X246. Z-0.01 F1000.
N66 Z-202.861
N67 X252. Z-203.153
N68 X254.828 Z-201.739
N69 G0 Z1.404
N70 X244.828
N71 G1 X242. Z-0.01 F1000.
N72 Z-202.667
```

```
N73 X248. Z-202.959
N74 X250.828 Z-201.545
N75 G0 Z1.404
N76 X240.828
N77 G1 X238. Z-0.01 F1000.
N78 Z-202.472
N79 X244. Z-202.764
N80 X246.828 Z-201.35
N81 G0 Z1.404
N82 X236.828
N83 G1 X234. Z-0.01 F1000.
N84 Z-202.277
N85 X240. Z-202.569
N86 X242.828 Z-201.155
N87 G0 Z1.404
N88 X232.828
N89 G1 X230. Z-0.01 F1000.
N90 Z-202.083
N91 X236. Z-202.375
N92 X238.828 Z-200.961
N93 G0 Z1.404
N94 X228.828
N95 G1 X226. Z-0.01 F1000.
N96 Z-201.888
N97 X232. Z-202.18
N98 X234.828 Z-200.766
N99 G0 Z1.404
N100 X224.828
N101 G1 X222. Z-0.01
F1000.
N102 Z-201.694
N103 X228. Z-201.985
N104 X230.828 Z-200.571
N105 G0 Z1.404
N106 X220.828
N107 G1 X218. Z-0.01
```

```
F1000.
N108 Z-201.499
N109 X224. Z-201.791
N110 X226.828 Z-200.377
N111 G0 Z1.404
N112 X216.828
N113 G1 X214. Z-0.01
F1000.
N114 Z-201.304
N115 X220. Z-201.596
N116 X222.828 Z-200.182
N117 G0 Z1.404
N118 X212.828
N119 G1 X210. Z-0.01
F1000.
N120 Z-201.11
N121 X216. Z-201.402
N122 X218.828 Z-199.987
N123 G0 Z1.404
N124 X208.828
N125 G1 X206. Z-0.01
F1000.
N126 Z-200.915
N127 X212. Z-201.207
N128 X214.828 Z-199.793
N129 G0 Z1.404
N130 X204.828
N131 G1 X202. Z-0.01
F1000.
N132 Z-200.72
N133 X208. Z-201.012
N134 X210.828 Z-199.598
N135 G0 Z1.404
N136 X200.828
N137 G1 X198. Z-0.01
F1000.
N138 Z-200.526
N139 X204. Z-200.818
N140 X206.828 Z-199.403
N141 G0 Z1.404
N142 X196.828
N143 G1 X194. Z-0.01
F1000.
N144 Z-200.331
N145 X200. Z-200.623
N146 X202.828 Z-199.209
N147 G0 Z1.404
N148 X192.828
N149 G1 X190. Z-0.01
F1000.
N150 Z-200.136
N151 X196. Z-200.428
N152 X198.828 Z-199.014
N153 G0 Z1.404
N154 X188.828
N155 G1 X186. Z-0.01
F1000.
N156 Z-199.942
N157 X192. Z-200.234
N158 X194.828 Z-198.819
N159 G0 Z1.404
N160 X184.828
N161 G1 X182. Z-0.01
F1000.
N162 Z-199.747
N163 X188. Z-200.039
N164 X190.828 Z-198.625
N165 G0 Z1.404
N166 X180.828
N167 G1 X178. Z-0.01
F1000.
N168 Z-199.552
N169 X184. Z-199.844
N170 X186.828 Z-198.43
N171 G0 Z1.404
N172 X176.828
N173 G1 X174. Z-0.01
F1000.
N174 Z-199.358
N175 X180. Z-199.65
N176 X182.828 Z-198.235
N177 G0 Z1.404
N178 X172.828
N179 G1 X170. Z-0.01
F1000.
N180 Z-3.324
N181 G3 X172.137 Z-4.969
R1.8
N182 G1 Z-152.672
N183 G3 X171.78 Z-156.754
R46.832
N184 G1 X170. Z-166.929
N185 Z-199.163
N186 X176. Z-199.455
N187 X178.828 Z-198.041
N188 G0 Z1.404
N189 X168.828
N190 G1 X166. Z-0.01
F1000.
N191 Z-3.169
N192 X168.537
N193 G3 X172. Z-4.477 R1.8
N194 G1 X174.828 Z-3.063
N195 G0 Z1.404
N196 X164.828
N197 G1 X162. Z-0.01
F1000.
N198 Z-3.169
N199 X168.
N200 X170.828 Z-1.755
N201 G0 Z1.404
N202 X160.828
N203 G1 X158. Z-0.01
F1000.
N204 Z-3.169
N205 X164.
N206 X166.828 Z-1.755
N207 G0 Z1.404
N208 X156.828
N209 G1 X154. Z-0.01
F1000.
N210 Z-3.169
N211 X160.
N212 X162.828 Z-1.755
N213 G0 Z1.404
N214 X152.828
N215 G1 X150. Z-0.01
F1000.
N216 Z-3.169
N217 X156.
N218 X158.828 Z-1.755
N219 G0 Z1.404
N220 X148.828
N221 G1 X146. Z-0.01
F1000.
N222 Z-3.169
N223 X152.
N224 X154.828 Z-1.755
N225 G0 Z1.404
N226 X144.828
N227 G1 X142. Z-0.01
F1000.
N228 Z-3.169
N229 X148.
N230 X150.828 Z-1.755
N231 G0 Z1.404
N232 X140.828
N233 G1 X138. Z-0.01
F1000.
N234 Z-3.169
N235 X144.
N236 X146.828 Z-1.755
N237 G0 Z1.404
N238 X136.828
N239 G1 X134. Z-0.01
F1000.
N240 Z-3.169
N241 X140.
N242 X142.828 Z-1.755
N243 G0 Z1.404
N244 X132.828
N245 G1 X130. Z-0.01
F1000.
N246 Z-3.169
N247 X136.
N248 X138.828 Z-1.755
N249 G0 Z1.404
N250 X128.828
N251 G1 X126. Z-0.01
F1000.
N252 Z-3.169
N253 X132.
N254 X134.828 Z-1.755
N255 G0 Z1.404
N256 X124.828
N257 G1 X122. Z-0.01
F1000.
N258 Z-3.169
N259 X128.
N260 X130.828 Z-1.755
N261 G0 Z1.404
N262 X120.828
N263 G1 X118. Z-0.01
F1000.
N264 Z-3.169
N265 X124.
N266 X126.828 Z-1.755
N267 G0 Z1.404
N268 X116.828
N269 G1 X114. Z-0.01
F1000.
N270 Z-3.169
N271 X120.
N272 X122.828 Z-1.755
N273 G0 Z1.404
N274 X112.828
N275 G1 X110. Z-0.01
F1000.
N276 Z-3.169
N277 X116.
N278 X118.828 Z-1.755
N279 G0 Z1.404
N280 X108.828
N281 G1 X106. Z-0.01
F1000.
N282 Z-3.169
N283 X112.
N284 X114.828 Z-1.755
N285 G0 Z1.404
N286 X104.828
N287 G1 X102. Z-0.01
F1000.
N288 Z-3.169
N289 X108.
N290 X110.828 Z-1.755
N291 G0 Z1.404
N292 X100.828
N293 G1 X98. Z-0.01 F1000.
N294 Z-3.169
```

```
N295 X104.                    N368 X58.828 Z-1.755           N441 G0 Z1.404
N296 X106.828 Z-1.755         N369 G0 Z1.404                 N442 X2.828
N297 G0 Z1.404                N370 X48.828                   N443 G1 X0. Z-0.01 F1000.
N298 X96.828                  N371 G1 X46. Z-0.01 F1000.     N444 Z-3.169
N299 G1 X94. Z-0.01 F1000.    N372 Z-3.169                   N445 X5.
N300 Z-3.169                  N373 X52.                      N446 X7.828 Z-1.755
N301 X100.                    N374 X54.828 Z-1.755           N447 X172.587
N302 X102.828 Z-1.755         N375 G0 Z1.404                 N448 G0 X172.828
N303 G0 Z1.404                N376 X44.828                   N449 Z-165.515
N304 X92.828                  N377 G1 X42. Z-0.01 F1000.     N450 G1 X170. Z-166.929
N305 G1 X90. Z-0.01 F1000.    N378 Z-3.169                   F1000.
N306 Z-3.169                  N379 X48.                      N451 X167.204 Z-182.91
N307 X96.                     N380 X50.828 Z-1.755           N452 Z-199.027
N308 X98.828 Z-1.755          N381 G0 Z1.404                 N453 X172. Z-199.26
N309 G0 Z1.404                N382 X40.828                   N454 X174.828 Z-197.846
N310 X88.828                  N383 G1 X38. Z-0.01 F1000.     N455 G0 Z-181.496
N311 G1 X86. Z-0.01 F1000.    N384 Z-3.169                   N456 X171.466
N312 Z-3.169                  N385 X44.                      N457 G1 X170.032 F1000.
N313 X92.                     N386 X46.828 Z-1.755           N458 X167.204 Z-182.91
N314 X94.828 Z-1.755          N387 G0 Z1.404                 N459 X164.407 Z-198.891
N315 G0 Z1.404                N388 X36.828                   N460 X169.204 Z-199.124
N316 X84.828                  N389 G1 X34. Z-0.01 F1000.     N461 X172.032 Z-197.71
N317 G1 X82. Z-0.01 F1000.    N390 Z-3.169                   N462 X181.436
N318 Z-3.169                  N391 X40.                      N463 G0 X256.828
N319 X88.                     N392 X42.828 Z-1.755           N464 Z-270.972
N320 X90.828 Z-1.755          N393 G0 Z1.404                 N465 G1 X254. Z-272.386
N321 G0 Z1.404                N394 X32.828                   F1000.
N322 X80.828                  N395 G1 X30. Z-0.01 F1000.     N466 X250.886 Z-290.181
N323 G1 X78. Z-0.01 F1000.    N396 Z-3.169                   N467 Z-309.135
N324 Z-3.169                  N397 X36.                      N468 X256. Z-311.038
N325 X84.                     N398 X38.828 Z-1.755           N469 X258.828 Z-309.624
N326 X86.828 Z-1.755          N399 G0 Z1.404                 N470 G0 Z-288.767
N327 G0 Z1.404                N400 X28.828                   N471 X255.149
N328 X76.828                  N401 G1 X26. Z-0.01 F1000.     N472 G1 X253.715 F1000.
N329 G1 X74. Z-0.01 F1000.    N402 Z-3.169                   N473 X250.886 Z-290.181
N330 Z-3.169                  N403 X32.                      N474 X247.772 Z-307.977
N331 X80.                     N404 X34.828 Z-1.755           N475 X252.886 Z-309.879
N332 X82.828 Z-1.755          N405 G0 Z1.404                 N476 X255.715 Z-308.465
N333 G0 Z1.404                N406 X24.828                   N477 G0 X255.946
N334 X72.828                  N407 G1 X22. Z-0.01 F1000.     N478 Z-2.755
N335 G1 X70. Z-0.01 F1000.    N408 Z-3.169                   N479 X171.961
N336 Z-3.169                  N409 X28.                      N480 G1 X1.228 F1000.
N337 X76.                     N410 X30.828 Z-1.755           N481 X-1.6 Z-4.169
N338 X78.828 Z-1.755          N411 G0 Z1.404                 N482 X168.537
N339 G0 Z1.404                N412 X20.828                   N483 G3 X170.137 Z-4.969
N340 X68.828                  N413 G1 X18. Z-0.01 F1000.     R0.8
N341 G1 X66. Z-0.01 F1000.    N414 Z-3.169                   N484 G1 Z-152.672
N342 Z-3.169                  N415 X24.                      N485 G3 X169.788 Z-156.667
N343 X72.                     N416 X26.828 Z-1.755           R45.832
N344 X74.828 Z-1.755          N417 G0 Z1.404                 N486 G1 X162.242 Z-199.79
N345 G0 Z1.404                N418 X16.828                   N487 X252.301 Z-204.173
N346 X64.828                  N419 G1 X14. Z-0.01 F1000.     N488 G3 X253.746 Z-204.969
N347 G1 X62. Z-0.01 F1000.    N420 Z-3.169                   R0.8
N348 Z-3.169                  N421 X20.                      N489 G1 Z-262.329
N349 X68.                     N422 X22.828 Z-1.755           N490 G3 X253.74 Z-262.399
N350 X70.828 Z-1.755          N423 G0 Z1.404                 R0.8
N351 G0 Z1.404                N424 X12.828                   N491 G1 X245.683 Z-308.446
N352 X60.828                  N425 G1 X10. Z-0.01 F1000.     N492 X265.155 Z-315.691
N353 G1 X58. Z-0.01 F1000.    N426 Z-3.169                   N493 G3 X265.8 Z-316.333
N354 Z-3.169                  N427 X16.                      R0.8
N355 X64.                     N428 X18.828 Z-1.755           N494 G1 Z-366.631
N356 X66.828 Z-1.755          N429 G0 Z1.404                 N495 X268.628 Z-365.217
N357 G0 Z1.404                N430 X8.828                    N496 X269.8
N358 X56.828                  N431 G1 X6. Z-0.01 F1000.      N497 G0 X290.
N359 G1 X54. Z-0.01 F1000.    N432 Z-3.169                   N498 Z5.
N360 Z-3.169                  N433 X12.                      N499 G53 X0.
N361 X60.                     N434 X14.828 Z-1.755
N362 X62.828 Z-1.755          N435 G0 Z1.404                 (Groove1)
N363 G0 Z1.404                N436 X5.828                    N500 M9
N364 X52.828                  N437 G1 X3. Z-0.01 F1000.      N501 M1
N365 G1 X50. Z-0.01 F1000.    N438 Z-3.169                   N502 T606
N366 Z-3.169                  N439 X8.                       N503 G98
N367 X56.                     N440 X10.828 Z-1.755           N504 M22
```

156

```
N505 G97 S500 M3          N578 G0 X170.21              N651 X141.998
N506 G54                  N579 Z-179.515               N652 X146.123
N507 M8                   N580 G1 X166.21 F1000.       N653 X148.123 Z-188.488
N508 G0 X290. Z5.         N581 X158.368                N654 X164.189
N509 G0 Z-169.543         N582 X160.665                N655 G0 X168.29
N510 X171.97              N583 X162.665 Z-178.515      N656 Z-190.485
N511 G1 X171.955 F1000.   N584 X165.934                N657 G1 X164.29 F1000.
N512 X167.955             N585 G0 X170.035             N658 X139.766
N513 X167.554             N586 Z-180.512               N659 X144.077
N514 X171.754             N587 G1 X166.035 F1000.      N660 X146.077 Z-189.485
N515 G0 X171.781          N588 X157.113                N661 X164.014
N516 Z-170.54             N589 X159.553                N662 G0 X168.116
N517 G1 X167.781 F1000.   N590 X161.553 Z-179.512      N663 Z-191.482
N518 X166.881             N591 X165.759                N664 G1 X164.116 F1000.
N519 X167.728             N592 G0 X169.861             N665 X137.383
N520 X167.781             N593 Z-181.51                N666 X142.082
N521 X171.554             N594 G1 X165.861 F1000.      N667 X144.082 Z-190.482
N522 G0 X171.606          N595 X155.786                N668 X163.84
N523 Z-171.537            N596 X158.383                N669 G0 X167.941
N524 G1 X167.606 F1000.   N597 X160.383 Z-180.51       N670 Z-192.479
N525 X166.158             N598 X165.585                N671 G1 X163.941 F1000.
N526 X167.538             N599 G0 X169.686             N672 X134.829
N527 X167.606             N600 Z-182.507               N673 X139.832
N528 X171.02              N601 G1 X165.686 F1000.      N674 X141.832 Z-191.479
N529 G0 X171.432          N602 X154.383                N675 X163.665
N530 Z-172.534            N603 X157.121                N676 G0 X167.767
N531 G1 X167.432 F1000.   N604 X159.121 Z-181.507      N677 Z-193.476
N532 X165.382             N605 X165.41                 N678 G1 X163.767 F1000.
N533 X166.863             N606 G0 X169.512             N679 X132.078
N534 X167.432             N607 Z-183.504               N680 X137.401
N535 X170.505             N608 G1 X165.512 F1000.      N681 X139.401 Z-192.476
N536 G0 X171.257          N609 X152.901                N682 X163.491
N537 Z-173.532            N610 X155.808                N683 G0 X167.592
N538 G1 X167.257 F1000.   N611 X157.808 Z-182.504      N684 Z-194.474
N539 X164.554             N612 X165.236                N685 G1 X163.592 F1000.
N540 X166.138             N613 G0 X169.337             N686 X129.101
N541 X167.257             N614 Z-184.501               N687 X134.894
N542 X170.332             N615 G1 X165.337 F1000.      N688 X136.894 Z-193.474
N543 G0 X171.083          N616 X151.334                N689 X163.316
N544 Z-174.529            N617 X154.397                N690 G0 X167.418
N545 G1 X167.083 F1000.   N618 X156.397 Z-183.501      N691 Z-195.471
N546 X163.671             N619 X165.061                N692 G1 X163.418 F1000.
N547 X165.364             N620 G0 X169.163             N693 X125.853
N548 X167.083             N621 Z-185.499               N694 X132.163
N549 X170.157             N622 G1 X165.163 F1000.      N695 X134.163 Z-194.471
N550 G0 X170.908          N623 X149.677                N696 X163.142
N551 Z-175.526            N624 X152.883                N697 G0 X167.243
N552 G1 X166.908 F1000.   N625 X154.883 Z-184.499      N698 Z-196.468
N553 X162.732             N626 X164.887                N699 G1 X163.243 F1000.
N554 X164.543             N627 G0 X168.988             N700 X122.275
N555 X166.908             N628 Z-186.496               N701 X129.156
N556 X169.981             N629 G1 X164.988 F1000.      N702 X131.156 Z-195.468
N557 G0 X170.733          N630 X147.923                N703 X163.032
N558 Z-176.523            N631 X151.323                N704 G0 X167.069
N559 G1 X166.733 F1000.   N632 X153.323 Z-185.496      N705 Z-197.465
N560 X161.734             N633 X164.712                N706 G1 X163.069 F1000.
N561 X163.667             N634 G0 X168.814             N707 X135.115
N562 X165.667 Z-175.523   N635 Z-187.493               N708 X137.115 Z-196.465
N563 X168.023             N636 G1 X164.814 F1000.      N709 X162.86
N564 G0 X170.559          N637 X146.064                N710 G0 X176.255
N565 Z-177.521            N638 X149.727                N711 Z-198.463
N566 G1 X166.559 F1000.   N639 X151.727 Z-186.493      N712 G1 X162.894 F1000.
N567 X160.676             N640 X164.538                N713 X155.608
N568 X162.721             N641 G0 X168.639             N714 X157.608 Z-197.463
N569 X164.721 Z-176.521   N642 Z-188.49                N715 X166.141
N570 X167.14              N643 G1 X164.639 F1000.      N716 G0 X257.748
N571 G0 X170.384          N644 X144.093                N717 Z-267.273
N572 Z-178.518            N645 X147.945                N718 X257.307
N573 G1 X166.384 F1000.   N646 X149.945 Z-187.49       N719 G1 X253.307 F1000.
N574 X159.554             N647 X164.363                N720 X252.848
N575 X161.716             N648 G0 X168.465             N721 X253.255
N576 X163.716 Z-177.518   N649 Z-189.488               N722 X253.307
N577 X166.202             N650 G1 X164.465 F1000.      N723 X257.106
```

```
N724 G0 X257.133          N797 X236.521              N870 X249.367
N725 Z-268.27             N798 X239.483              N871 G0 X253.468
N726 G1 X253.132 F1000.   N799 X241.483 Z-277.242    N872 Z-289.212
N727 X251.363             N800 X251.111              N873 G1 X249.468 F1000.
N728 X253.084             N801 G0 X255.213           N874 X222.759
N729 X253.132             N802 Z-279.24              N875 X224.324
N730 X256.848             N803 G1 X251.213 F1000.    N876 X226.324 Z-288.212
N731 G0 X256.958          N804 X235.037              N877 X249.192
N732 Z-269.267            N805 X238.007              N878 G0 X253.294
N733 G1 X252.958 F1000.   N806 X240.007 Z-278.24     N879 Z-290.209
N734 X249.879             N807 X250.937              N880 G1 X249.294 F1000.
N735 X252.838             N808 G0 X255.039           N881 X222.153
N736 X252.958             N809 Z-280.237             N882 X223.531
N737 X256.314             N810 G1 X251.039 F1000.    N883 X225.531 Z-289.209
N738 G0 X256.784          N811 X233.553              N884 X249.018
N739 Z-270.264            N812 X236.558              N885 G0 X253.119
N740 G1 X252.784 F1000.   N813 X238.558 Z-279.237    N886 Z-291.206
N741 X248.395             N814 X250.762              N887 G1 X249.119 F1000.
N742 X251.344             N815 G0 X254.864           N888 X221.656
N743 X252.784             N816 Z-281.234             N889 X222.836
N744 X255.858             N817 G1 X250.864 F1000.    N890 X224.836 Z-290.206
N745 G0 X256.609          N818 X232.068              N891 X248.843
N746 Z-271.262            N819 X235.079              N892 G0 X252.945
N747 G1 X252.609 F1000.   N820 X237.079 Z-280.234    N893 Z-292.204
N748 X246.911             N821 X250.588              N894 G1 X248.945 F1000.
N749 X249.871             N822 G0 X254.69            N895 X221.264
N750 X252.609             N823 Z-282.231             N896 X222.237
N751 X255.684             N824 G1 X250.69 F1000.     N897 X224.237 Z-291.204
N752 G0 X256.435          N825 X230.584              N898 X248.669
N753 Z-272.259            N826 X233.569              N899 G0 X252.77
N754 G1 X252.435 F1000.   N827 X235.569 Z-281.231    N900 Z-293.201
N755 X245.427             N828 X250.413              N901 G1 X248.77 F1000.
N756 X248.383             N829 G0 X254.515           N902 X220.975
N757 X250.383 Z-271.259   N830 Z-283.229             N903 X221.735
N758 X253.348             N831 G1 X250.515 F1000.    N904 X223.735 Z-292.201
N759 G0 X256.26           N832 X229.1                N905 X248.494
N760 Z-273.256            N833 X232.111              N906 G0 X252.596
N761 G1 X252.26 F1000.    N834 X234.111 Z-282.229    N907 Z-294.198
N762 X243.942             N835 X250.239              N908 G1 X248.596 F1000.
N763 X246.899             N836 G0 X254.341           N909 X220.785
N764 X248.899 Z-272.256   N837 Z-284.226             N910 X221.328
N765 X251.984             N838 G1 X250.341 F1000.    N911 X223.328 Z-293.198
N766 G0 X256.086          N839 X227.668              N912 X248.32
N767 Z-274.253            N840 X230.591              N913 G0 X252.421
N768 G1 X252.086 F1000.   N841 X232.591 Z-283.226    N914 Z-295.195
N769 X242.458             N842 X250.064              N915 G1 X248.421 F1000.
N770 X245.429             N843 G0 X254.166           N916 X220.693
N771 X247.429 Z-273.253   N844 Z-285.223             N917 X221.018
N772 X251.809             N845 G1 X250.166 F1000.    N918 X223.018 Z-294.195
N773 G0 X255.911          N846 X226.403              N919 X248.21
N774 Z-275.251            N847 X229.095              N920 G0 X252.247
N775 G1 X251.911 F1000.   N848 X231.095 Z-284.223    N921 Z-296.193
N776 X240.974             N849 X249.89               N922 G1 X248.247 F1000.
N777 X243.964             N850 G0 X253.992           N923 X220.793
N778 X245.964 Z-274.251   N851 Z-286.22              N924 X222.793 Z-295.193
N779 X251.635             N852 G1 X249.992 F1000.    N925 X248.036
N780 G0 X255.737          N853 X225.292              N926 G0 X252.073
N781 Z-276.248            N854 X227.704              N927 Z-297.19
N782 G1 X251.737 F1000.   N855 X229.704 Z-285.22     N928 G1 X248.073 F1000.
N783 X239.49              N856 X249.715              N929 X221.317
N784 X242.456             N857 G0 X253.817           N930 X223.317 Z-296.19
N785 X244.456 Z-275.248   N858 Z-287.217             N931 X247.861
N786 X251.46              N859 G1 X249.817 F1000.    N932 G0 X251.898
N787 G0 X255.562          N860 X224.321              N933 Z-298.187
N788 Z-277.245            N861 X226.412              N934 G1 X247.898 F1000.
N789 G1 X251.562 F1000.   N862 X228.412 Z-286.217    N935 X222.356
N790 X238.005             N863 X249.541              N936 X224.356 Z-297.187
N791 X240.971             N864 G0 X253.643           N937 X247.687
N792 X242.971 Z-276.245   N865 Z-288.215             N938 G0 X251.724
N793 X251.286             N866 G1 X249.643 F1000.    N939 Z-299.184
N794 G0 X255.388          N867 X223.479              N940 G1 X247.724 F1000.
N795 Z-278.242            N868 X225.319              N941 X224.167
N796 G1 X251.388 F1000.   N869 X227.319 Z-287.215    N942 X226.167 Z-298.184
```

```
N943 X247.513
N944 G0 X251.549
N945 Z-300.182
N946 G1 X247.549 F1000.
N947 X226.823
N948 X228.823 Z-299.182
N949 X247.338
N950 G0 X251.375
N951 Z-301.179
N952 G1 X247.375 F1000.
N953 X229.503
N954 X231.503 Z-300.179
N955 X247.163
N956 G0 X251.2
N957 Z-302.176
N958 G1 X247.2 F1000.
N959 X232.183
N960 X234.183 Z-301.176
N961 X246.989
N962 G0 X251.026
N963 Z-303.173
N964 G1 X247.026 F1000.
N965 X234.863
N966 X236.863 Z-302.173
N967 X246.814
N968 G0 X250.851
N969 Z-304.171
N970 G1 X246.851 F1000.
N971 X237.544
N972 X239.544 Z-303.171
N973 X246.639
N974 G0 X250.676
N975 Z-305.168
N976 G1 X246.676 F1000.
N977 X240.224
N978 X242.224 Z-304.168
N979 X246.457
N980 G0 X250.502
N981 Z-306.165
N982 G1 X246.502 F1000.
N983 X242.904
N984 X244.904 Z-305.165
N985 X249.046
N986 G0 X250.327
N987 Z-307.162
N988 G1 X246.327 F1000.
N989 X245.584
N990 X246.327
N991 X250.089
N992 G0 X257.728
N993 Z-167.238
N994 X172.138

N995 G1 X170.941 F1000.
N996 X168.113 Z-168.652
N997 G18 G3 X122.235 Z-
196.479 R46.833
N998 G1 Z-196.659
N999 X125.666 Z-195.631
N1000 X157.195
N1001 G0 X166.828
N1002 Z-198.912
N1003 G1 X162.833 Z-
198.814 F1000.
N1004 X122.235 Z-196.839
N1005 Z-196.659
N1006 X126.192 Z-196.952
N1007 X164.287
N1008 G0 X173.992
N1009 Z-157.482
N1010 G1 X172.646 F1000.
N1011 X169.817 Z-158.896
N1012 G3 X120.235 Z-
195.871 R45.832
N1013 G1 Z-197.746
N1014 X252.129 Z-204.164
N1015 X254.957 Z-202.75
N1016 X255.527
N1017 G0 X257.728
N1018 Z-265.509
N1019 X257.503
N1020 G1 X256.197 F1000.
N1021 X253.368 Z-266.923
N1022 X228.738 Z-283.472
N1023 G2 X220.685 Z-
295.522 R20.364
N1024 G1 X223.513 Z-
294.108
N1025 X224.812
N1026 G0 X250.102
N1027 Z-308.613
N1028 G1 X249.494 F1000.
N1029 X246.282 Z-307.422
N1030 X224.577 Z-299.346
N1031 G3 X220.685 Z-
295.522 R4.831
N1032 G1 X223.513 Z-
296.937
N1033 X225.504
N1034 G0 X257.728
N1035 Z-263.388
N1036 X257.726
N1037 G1 X256.568 F1000.
N1038 X253.739 Z-264.802
N1039 G3 X253.43 Z-265.207

R0.8
N1040 G1 X227.134 Z-
282.875
N1041 G2 X218.685 Z-
295.525 R21.364
N1042 X223.383 Z-300.148
R5.831
N1043 G1 X246.077 Z-
308.592
N1044 X248.905 Z-307.178
N1045 X248.942
N1046 G0 X257.741
N1047 Z-4.169
N1048 X173.882
N1049 G1 X-0.02 F1000.
N1050 X168.542
N1051 X170.542 Z-3.169
N1052 X172.798
N1053 G0 X174.119
N1054 Z-156.471
N1055 X174.074
N1056 G1 X172.792 F1000.
N1057 X169.964 Z-157.885
N1058 X169.963 Z-157.892
N1059 X172.792 Z-156.478
N1060 G0 Z-157.024
N1061 G1 X172.718 F1000.
N1062 X169.889 Z-158.438
N1063 G3 X169.884 Z-
158.471 R45.832
N1064 G1 X172.713 Z-
157.056
N1065 X174.033
N1066 G0 X257.728
N1067 Z-263.315
N1068 X257.726
N1069 G1 X256.574 F1000.
N1070 X253.746 Z-264.729
N1071 G3 X253.739 Z-
264.802 R0.8
N1072 G1 X256.568 Z-
263.388
N1073 X257.746
N1074 G0 X290.
N1075 Z5.

N1076 M9
N1077 G53 X0.
N1078 G53 Z0.
N1079 M30
%
```

Setup Sheet - Part 9004

Job

WCS: #0

STOCK:

 DX: 270mm
DY: 270mm
DZ: 370mm

PART:

 DX: 265.8mm
 DY: 265.8mm
 DZ: 361.66mm

STOCK LOWER IN WCS #0:
 X: -135mm

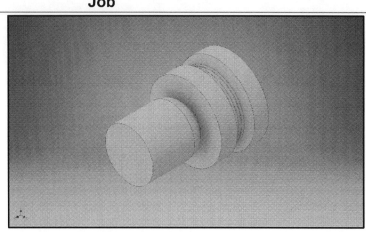

Total

NUMBER OF OPERATIONS: 1
NUMBER OF TOOLS: 1
TOOLS: T3

MAXIMUM Z: 5.8mm

MINIMUM Z: -365.83mm

MAXIMUM FEEDRATE: 1000mm/min
MAXIMUM SPINDLE SPEED: 500rpm

Operation 1/1

DESCRIPTION: Profile1
STRATEGY: Unspecified
WCS: #0

TOLERANCE: 0.01mm
STOCK TO LEAVE: 0mm

MAXIMUM Z: 5.8mm

MINIMUM Z: -365.83mm

MAXIMUM SPINDLE SPEED: 500rpm
MAXIMUM FEEDRATE: 1000mm/min
CUTTING DISTANCE: 11932.7mm
RAPID DISTANCE: 12216.74mm

T3 D0 L0

TYPE: general turning

DIAMETER: 0mm

Generated by Inventor HSM Pro 4.0.0.032

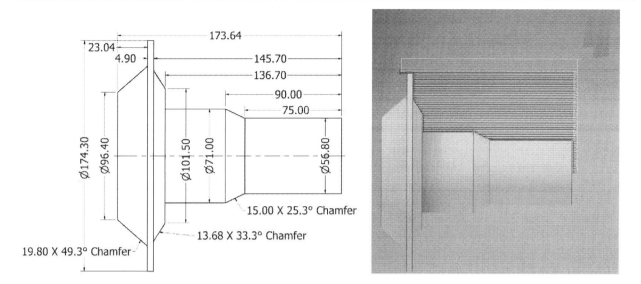

Figure 5.10 Drawing and tool path for part 1101

```
%
O1101 (Part 1101)
N10 G98 G18
N11 G21
N12 G50 S6000
N13 M31
N14 G53 G0 X0.

(Profile1)
N15 T101
N16 G98
N17 M22
N18 G97 S500 M3
N19 G54
N20 M8
N21 G0 X200. Z5.
N22 G0 Z1.404
N23 X181.821
N24 G1 X178.828 F1000.
N25 X176. Z-0.01
N26 Z-148.959
N27 G18 G3 X176.299 Z-
149.678 R1.8
N28 G1 Z-154.578
N29 G3 X176.286 Z-154.735
R1.8
N30 G1 X176.038 Z-156.148
N31 Z-160.622
N32 X179.98
N33 X182.808 Z-159.208
N34 G0 Z1.404
N35 X174.828
N36 G1 X172. Z-0.01 F1000.
N37 Z-147.878
N38 X172.699
N39 G3 X176.299 Z-149.678
R1.8
N40 G1 X179.128 Z-148.264
N41 G0 Z1.404
```

```
N42 X170.828
N43 G1 X168. Z-0.01 F1000.
N44 Z-147.878
N45 X172.699
N46 G3 X174. Z-148. R1.8
N47 G1 X176.828 Z-146.585
N48 G0 Z1.404
N49 X166.828
N50 G1 X164. Z-0.01 F1000.
N51 Z-147.878
N52 X170.
N53 X172.828 Z-146.464
N54 G0 Z1.404
N55 X162.828
N56 G1 X160. Z-0.01 F1000.
N57 Z-147.878
N58 X166.
N59 X168.828 Z-146.464
N60 G0 Z1.404
N61 X158.828
N62 G1 X156. Z-0.01 F1000.
N63 Z-147.878
N64 X162.
N65 X164.828 Z-146.464
N66 G0 Z1.404
N67 X154.828
N68 G1 X152. Z-0.01 F1000.
N69 Z-147.878
N70 X158.
N71 X160.828 Z-146.464
N72 G0 Z1.404
N73 X150.828
N74 G1 X148. Z-0.01 F1000.
N75 Z-147.878
N76 X154.
N77 X156.828 Z-146.464
N78 G0 Z1.404
N79 X146.828
N80 G1 X144. Z-0.01 F1000.
N81 Z-147.878
```

```
N82 X150.
N83 X152.828 Z-146.464
N84 G0 Z1.404
N85 X142.828
N86 G1 X140. Z-0.01 F1000.
N87 Z-147.878
N88 X146.
N89 X148.828 Z-146.464
N90 G0 Z1.404
N91 X138.828
N92 G1 X136. Z-0.01 F1000.
N93 Z-147.878
N94 X142.
N95 X144.828 Z-146.464
N96 G0 Z1.404
N97 X134.828
N98 G1 X132. Z-0.01 F1000.
N99 Z-147.878
N100 X138.
N101 X140.828 Z-146.464
N102 G0 Z1.404
N103 X130.828
N104 G1 X128. Z-0.01
F1000.
N105 Z-147.768
N106 X128.333 Z-147.878
N107 X134.
N108 X136.828 Z-146.464
N109 G0 Z1.404
N110 X126.828
N111 G1 X124. Z-0.01
F1000.
N112 Z-146.452
N113 X128.333 Z-147.878
N114 X130.
N115 X132.828 Z-146.464
N116 G0 Z1.404
N117 X122.828
N118 G1 X120. Z-0.01
F1000.
```

```
N119 Z-145.136              N185 G0 Z1.404               N258 G0 Z1.404
N120 X126. Z-147.11         N186 X78.828                 N259 X34.828
N121 X128.828 Z-145.696     N187 G1 X76. Z-0.01 F1000.   N260 G1 X32. Z-0.01 F1000.
N122 G0 Z1.404              N188 Z-138.878               N261 Z-2.178
N123 X118.828               N189 X82.                    N262 X38.
N124 G1 X116. Z-0.01        N190 X84.828 Z-137.464       N263 X40.828 Z-0.763
F1000.                      N191 G0 Z1.404               N264 G0 Z1.404
N125 Z-143.82               N192 X74.828                 N265 X30.828
N126 X122. Z-145.794        N193 G1 X72. Z-0.01 F1000.   N266 G1 X28. Z-0.01 F1000.
N127 X124.828 Z-144.38      N194 Z-92.517                N267 Z-2.178
N128 G0 Z1.404              N195 X72.654 Z-93.208        N268 X34.
N129 X114.828               N196 G3 X73. Z-93.978 R1.8   N269 X36.828 Z-0.763
N130 G1 X112. Z-0.01        N197 G1 Z-138.878            N270 G0 Z1.404
F1000.                      N198 X78.                    N271 X26.828
N131 Z-142.504              N199 X80.828 Z-137.464       N272 G1 X24. Z-0.01 F1000.
N132 X118. Z-144.478        N200 G0 Z1.404               N273 Z-2.178
N133 X120.828 Z-143.064     N201 X70.828                 N274 X30.
N134 G0 Z1.404              N202 G1 X68. Z-0.01 F1000.   N275 X32.828 Z-0.763
N135 X110.828               N203 Z-88.291                N276 G0 Z1.404
N136 G1 X108. Z-0.01        N204 X72.654 Z-93.208        N277 X22.828
F1000.                      N205 G3 X73. Z-93.978 R1.8   N278 G1 X20. Z-0.01 F1000.
N137 Z-141.188              N206 G1 X75.828 Z-92.563     N279 Z-2.178
N138 X114. Z-143.162        N207 G0 Z1.404               N280 X26.
N139 X116.828 Z-141.748     N208 X66.828                 N281 X28.828 Z-0.763
N140 G0 Z1.404              N209 G1 X64. Z-0.01 F1000.   N282 G0 Z1.404
N141 X106.828               N210 Z-84.066                N283 X18.828
N142 G1 X104. Z-0.01        N211 X70. Z-90.404           N284 G1 X16. Z-0.01 F1000.
F1000.                      N212 X72.828 Z-88.99         N285 Z-2.178
N143 Z-139.872              N213 G0 Z1.404               N286 X22.
N144 X110. Z-141.846        N214 X62.828                 N287 X24.828 Z-0.763
N145 X112.828 Z-140.432     N215 G1 X60. Z-0.01 F1000.   N288 G0 Z1.404
N146 G0 Z1.404              N216 Z-79.841                N289 X14.828
N147 X102.828               N217 X66. Z-86.179           N290 G1 X12. Z-0.01 F1000.
N148 G1 X100. Z-0.01        N218 X68.828 Z-84.765        N291 Z-2.178
F1000.                      N219 G0 Z1.404               N292 X18.
N149 Z-138.879              N220 X58.828                 N293 X20.828 Z-0.763
N150 G3 X101.879 Z-139.174  N221 G1 X56. Z-0.01 F1000.   N294 G0 Z1.404
R1.8                        N222 Z-2.223                 N295 X10.828
N151 G1 X106. Z-140.53      N223 G3 X58.8 Z-3.978 R1.8   N296 G1 X8. Z-0.01 F1000.
N152 X108.828 Z-139.116     N224 G1 Z-78.573             N297 Z-2.178
N153 G0 Z1.404              N225 X62. Z-81.953           N298 X14.
N154 X98.828               N226 X64.828 Z-80.539        N299 X16.828 Z-0.763
N155 G1 X96. Z-0.01 F1000.  N227 G0 Z1.404               N300 G0 Z1.404
N156 Z-138.878              N228 X54.828                 N301 X6.828
N157 X99.9                  N229 G1 X52. Z-0.01 F1000.   N302 G1 X4. Z-0.01 F1000.
N158 G3 X101.879 Z-139.174  N230 Z-2.178                 N303 Z-2.178
R1.8                        N231 X55.2                   N304 X10.
N159 G1 X102. Z-139.214     N232 G3 X58. Z-2.846 R1.8    N305 X12.828 Z-0.763
N160 X104.828 Z-137.8       N233 G1 X60.828 Z-1.432      N306 G0 Z1.404
N161 G0 Z1.404              N234 G0 Z1.404               N307 X2.828
N162 X94.828               N235 X50.828                 N308 G1 X0. Z-0.01 F1000.
N163 G1 X92. Z-0.01 F1000.  N236 G1 X48. Z-0.01 F1000.   N309 Z-2.178
N164 Z-138.878              N237 Z-2.178                 N310 X6.
N165 X98.                   N238 X54.                    N311 X8.828 Z-0.763
N166 X100.828 Z-137.464     N239 X56.828 Z-0.763         N312 X59.25
N167 G0 Z1.404              N240 G0 Z1.404               N313 G0 X178.867
N168 X90.828               N241 X46.828                 N314 Z-154.734
N169 G1 X88. Z-0.01 F1000.  N242 G1 X44. Z-0.01 F1000.   N315 G1 X176.038 Z-156.148
N170 Z-138.878              N243 Z-2.178                 F1000.
N171 X94.                   N244 X50.                    N316 X175.256 Z-160.622
N172 X96.828 Z-137.464      N245 X52.828 Z-0.763         N317 X176.038
N173 G0 Z1.404              N246 G0 Z1.404               N318 X178.867 Z-159.208
N174 X86.828               N247 X42.828                 N319 G0 Z-1.763
N175 G1 X84. Z-0.01 F1000.  N248 G1 X40. Z-0.01 F1000.   N320 X58.625
N176 Z-138.878              N249 Z-2.178                 N321 G1 X1.228 F1000.
N177 X90.                   N250 X46.                    N322 X-1.6 Z-3.178
N178 X92.828 Z-137.464      N251 X48.828 Z-0.763         N323 X55.2
N179 G0 Z1.404              N252 G0 Z1.404               N324 G3 X56.8 Z-3.978 R0.8
N180 X82.828               N253 X38.828                 N325 G1 Z-78.798
N181 G1 X80. Z-0.01 F1000.  N254 G1 X36. Z-0.01 F1000.   N326 X70.846 Z-93.635
N182 Z-138.878              N255 Z-2.178                 N327 G3 X71. Z-93.978 R0.8
N183 X86.                   N256 X42.                    N328 G1 Z-139.878
N184 X88.828 Z-137.464      N257 X44.828 Z-0.763         N329 X99.9
```

162

```
N330 G3 X100.779 Z-140.01        N335 G3 X174.293 Z-154.648
R0.8                             R0.8                          N341 M9
N331 G1 X127.734 Z-148.878       N336 G1 X173.248 Z-160.622    N342 G53 X0.
N332 X172.699                    N337 X176.076 Z-159.208       N343 G53 Z0.
N333 G3 X174.299 Z-149.678       N338 X177.511                 N344 M30
R0.8                             N339 G0 X200.                    %
N334 G1 Z-154.578                N340 Z5.
```

Setup Sheet – Part1101

Job

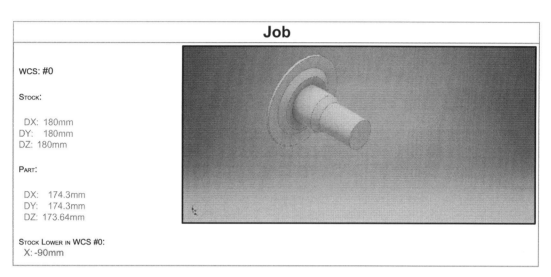

WCS: #0

STOCK:

 DX: 180mm
DY: 180mm
DZ: 180mm

PART:

 DX: 174.3mm
 DY: 174.3mm
 DZ: 173.64mm

STOCK LOWER IN WCS #0:
 X: -90mm

Total

NUMBER OF OPERATIONS: 1
NUMBER OF TOOLS: 1
TOOLS: T1

MAXIMUM Z: 5mm

MINIMUM Z: -160.62mm

MAXIMUM FEEDRATE: 1000mm/min
MAXIMUM SPINDLE SPEED: 500rpm

Operation 1/1		
	MAXIMUM Z: 5mm	T1 D0 L0
DESCRIPTION: Profile1		
STRATEGY: Unspecified	MINIMUM Z: -160.62mm	TYPE: general turning
WCS: #0		
	MAXIMUM SPINDLE SPEED: 500rpm	DIAMETER: 0mm
TOLERANCE: 0.01mm	MAXIMUM FEEDRATE: 1000mm/min	
STOCK TO LEAVE: 0mm	CUTTING DISTANCE: 4898.37mm	
	RAPID DISTANCE: 5138.54mm	

Generated by Inventor HSM Pro 4.0.0.032

163

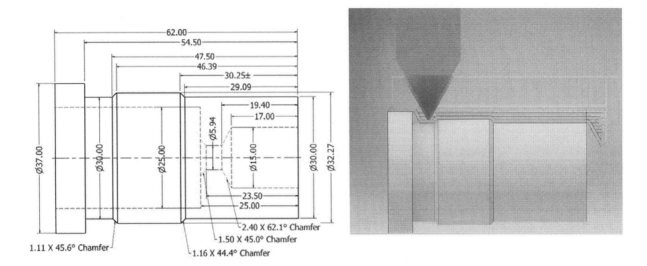

Figure 5.11 Drawing and tool path for part 9006

```
%
O9006 (Part 9006)
N10 G98 G18
N11 G21
N12 G50 S6000
N13 M31
N14 G53 G0 X0.

(Face2)
N15 T202
N16 G98
N17 M22
N18 G97 S500 M3
N19 G54
N20 M8
N21 G0 X58. Z5.
N22 G0 Z-2.586
N23 G1 X40.828 F1000.
N24 X38. Z-4.
N25 X-0.4
N26 X2.428 Z-2.586
N27 G0 X58.
N28 Z5.
N29 G53 X0.

(Profile4)
N30 M9
N31 M1
N32 T303
N33 G98
N34 M22
N35 G97 S500 M3
N36 G54
N37 M8
N38 G0 X58. Z5.8
N39 G0 Z2.204
N40 X39.821
N41 G1 X38.828 F1000.
```

```
N42 X36. Z0.79
N43 Z-56.595
N44 X37.2 Z-56.941
N45 G18 G3 X37.98 Z-57.245
R1.8
N46 G1 X40.809 Z-55.831
N47 G0 Z2.204
N48 X36.828
N49 G1 X34. Z0.79 F1000.
N50 Z-33.561
N51 G3 X34.271 Z-34.247
R1.8
N52 G1 Z-50.387
N53 G3 X34. Z-51.072 R1.8
N54 G1 Z-56.017
N55 X37. Z-56.883
N56 X39.828 Z-55.469
N57 G0 Z2.204
N58 X36.828
N59 G1 X32. Z0.79 F1000.
N60 Z-3.997
N61 Z-4.
N62 Z-32.353
N63 X33.242 Z-32.987
N64 G3 X34.271 Z-34.247
R1.8
N65 G1 X37.1 Z-32.832
N66 G0 Z2.204
N67 X32.828
N68 G1 X30. Z0.79 F1000.
N69 Z-2.383
N70 X30.2 Z-2.441
N71 G3 X32. Z-3.997 R1.8
N72 G1 Z-4.
N73 X34.828 Z-2.586
N74 G0 Z2.204
N75 X30.828
N76 G1 X28. Z0.79 F1000.
```

```
N77 Z-1.806
N78 X30.2 Z-2.441
N79 G3 X31. Z-2.755 R1.8
N80 G1 X33.828 Z-1.341
N81 G0 Z2.204
N82 X28.828
N83 G1 X26. Z0.79 F1000.
N84 Z-1.229
N85 X29. Z-2.095
N86 X31.828 Z-0.681
N87 G0 Z2.204
N88 X26.828
N89 G1 X24. Z0.79 F1000.
N90 Z-0.651
N91 X27. Z-1.517
N92 X29.828 Z-0.103
N93 G0 Z2.204
N94 X24.828
N95 G1 X22. Z0.79 F1000.
N96 Z-0.074
N97 X25. Z-0.94
N98 X27.828 Z0.474
N99 G0 Z2.204
N100 X23.315
N101 G1 X20.486 Z0.79
F1000.
N102 Z0.363
N103 X23. Z-0.363
N104 X25.828 Z1.052
N105 G0 Z2.204
N106 X22.108
N107 G1 X21.835 F1000.
N108 X19.007 Z0.79
N109 X21.486 Z0.074
N110 X24.315 Z1.489
N111 X24.587
N112 G0 X36.828
N113 Z-49.658
```

```
N114 G1 X34. Z-51.072
F1000.
N115 G3 X33.191 Z-51.672
R1.8
N116 G1 X32. Z-52.256
N117 Z-55.44
N118 X35. Z-56.306
N119 X37.828 Z-54.892
N120 G0 Z2.107
N121 X18.445
N122 G1 X18.172 F1000.
N123 X15.344 Z0.693
N124 X29.2 Z-3.307
N125 G3 X30. Z-4. R0.8
N126 G1 Z-32.761
N127 X31.814 Z-33.687
N128 G3 X32.271 Z-34.247
R0.8
N129 G1 Z-50.387
N130 G3 X31.791 Z-50.958
R0.8
N131 G1 X30. Z-51.836
N132 Z-56.017
N133 X36.2 Z-57.807
N134 G3 X37. Z-58.5 R0.8
N135 G1 Z-66.
N136 X39.828 Z-64.586
N137 X41.
N138 G0 X58.
N139 Z5.8

N140 M9
N141 G53 X0.
N142 G53 Z0.
N143 M30
%
```

Setup Sheet – Part 9006

Job

WCS: #0

STOCK:

 DX: 38mm
DY: 38mm
DZ: 70mm

PART:

 DX: 37mm
 DY: 37mm
 DZ: 62mm

STOCK LOWER IN WCS #0:
 X: -19mm

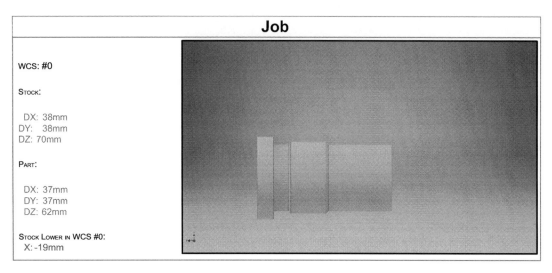

Total

NUMBER OF OPERATIONS: 2
NUMBER OF TOOLS: 2
TOOLS: **T2 T3**

MAXIMUM Z: 5.8mm

MINIMUM Z: -66mm

MAXIMUM FEEDRATE: 1000mm/min
MAXIMUM SPINDLE SPEED: 500rpm

Operation 1/2

DESCRIPTION: Face2	MAXIMUM Z: 5mm	**T2** D0 L0
STRATEGY: Unspecified	MINIMUM Z: –4mm	TYPE: general turning
WCS: #0	MAXIMUM SPINDLE SPEED: 500rpm	DIAMETER: 0mm
TOLERANCE: 0.01mm	MAXIMUM FEEDRATE: 1000mm/min	LENGTH: 0mm
	CUTTING DISTANCE: 31.79mm	FLUTES: 1

Operation 2/2

DESCRIPTION: Profile4	MAXIMUM Z: 5.8mm	**T3** D0 L0
STRATEGY: Unspecified	MINIMUM Z: -66mm	TYPE: general turning
WCS: #0	MAXIMUM SPINDLE SPEED: 500rpm	DIAMETER: 0mm
TOLERANCE: 0.01mm	MAXIMUM FEEDRATE: 1000mm/min	LENGTH: 0mm
STOCK TO LEAVE: 0mm	CUTTING DISTANCE: 304.57mm	FLUTES: 1

Generated by Inventor HSM Pro 4.0.0.032

Part 9007

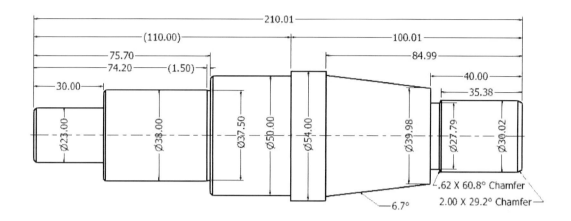

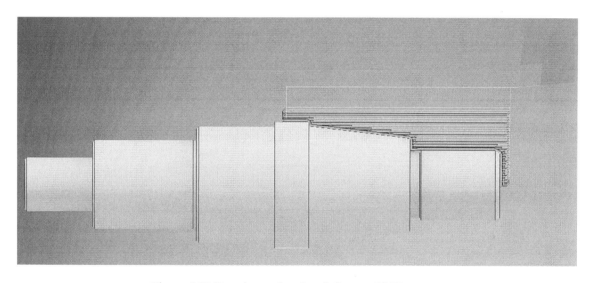

Figure 5.12 Drawing and tool path for part 9007

```
%
O9007 (PART 9007)
N10 G98 G18
N11 G21
N12 G50 S6000
N13 M31
N14 G53 G0 X0.

(Profile1)
N15 T101
N16 G98
N17 M22
N18 G97 S500 M3
N19 G54
N20 M8
N21 G0 X83.82 Z5.
N22 G0 Z1.427
N23 X65.367
N24 G1 X64.342 F1000.
N25 X61.468 Z-0.01
N26 Z-99.454
```

```
N27 X63.48
N28 X66.354 Z-98.017
N29 G0 Z1.427
N30 X62.31
N31 G1 X59.436 Z-0.01
F1000.
N32 Z-99.454
N33 X61.468
N34 X64.342 Z-98.017
N35 G0 Z1.427
N36 X60.278
N37 G1 X57.404 Z-0.01
F1000.
N38 Z-99.454
N39 X59.436
N40 X62.31 Z-98.017
N41 G0 Z1.427
N42 X58.246
N43 G1 X55.372 Z-0.01
F1000.
N44 Z-87.691
```

```
N45 G18 G3 X56.032 Z-88.735
R1.816
N46 G1 Z-99.454
N47 X57.404
N48 X60.278 Z-98.017
N49 G0 Z1.427
N50 X56.214
N51 G1 X53.34 Z-0.01 F1000.
N52 Z-86.981
N53 G3 X56.032 Z-88.735
R1.816
N54 G1 X58.906 Z-87.298
N55 G0 Z1.427
N56 X54.182
N57 G1 X51.308 Z-0.01
F1000.
N58 Z-83.221
N59 X52.176 Z-86.919
N60 X52.4
N61 G3 X54.356 Z-87.205
R1.816
```

```
N62 G1 X57.23 Z-85.768        N123 G1 X30.988 Z-0.01        N181 Z-1.93
N63 G0 Z1.427                 F1000.                        N182 X17.78
N64 X52.15                    N124 Z-4.319                  N183 X20.654 Z-0.494
N65 G1 X49.276 Z-0.01         N125 X31.593 Z-4.861          N184 G0 Z1.427
F1000.                        N126 G3 X32.055 Z-5.746       N185 X15.574
N66 Z-74.57                   R1.816                        N186 G1 X12.7 Z-0.01 F1000.
N67 X52.176 Z-86.919          N127 G1 Z-39.122              N187 Z-1.93
N68 X52.324                   N128 G3 X32.041 Z-39.28       N188 X15.748
N69 X55.198 Z-85.482          R1.816                        N189 X18.622 Z-0.494
N70 G0 Z1.427                 N129 G1 X31.577 Z-41.93       N190 G0 Z1.427
N71 X50.118                   N130 X34.036                  N191 X13.542
N72 G1 X47.244 Z-0.01         N131 X36.91 Z-40.494          N192 G1 X10.668 Z-0.01
F1000.                        N132 G0 Z1.427                F1000.
N73 Z-65.918                  N133 X31.83                   N193 Z-1.93
N74 X50.292 Z-78.895          N134 G1 X28.956 Z-0.01        N194 X13.716
N75 X53.166 Z-77.459          F1000.                        N195 X16.59 Z-0.494
N76 G0 Z1.427                 N135 Z-2.571                  N196 G0 Z1.427
N77 X48.086                   N136 G3 X29.358 Z-2.861       N197 X11.51
N78 G1 X45.212 Z-0.01         R1.816                        N198 G1 X8.636 Z-0.01
F1000.                        N137 G1 X31.593 Z-4.861       F1000.
N79 Z-57.266                  N138 G3 X32.004 Z-5.444       N199 Z-1.93
N80 X48.26 Z-70.244           R1.816                        N200 X11.684
N81 X51.134 Z-68.807          N139 G1 X34.878 Z-4.007       N201 X14.558 Z-0.494
N82 G0 Z1.427                 N140 G0 Z1.427                N202 G0 Z1.427
N83 X46.054                   N141 X29.798                  N203 X9.478
N84 G1 X43.18 Z-0.01 F1000.   N142 G1 X26.924 Z-0.01        N204 G1 X6.604 Z-0.01
N85 Z-48.615                  F1000.                        F1000.
N86 X46.228 Z-61.592          N143 Z-1.968                  N205 Z-1.93
N87 X49.102 Z-60.155          N144 G3 X29.358 Z-2.861       N206 X9.652
N88 G0 Z1.427                 R1.816                        N207 X12.526 Z-0.494
N89 X44.022                   N145 G1 X29.972 Z-3.41        N208 G0 Z1.427
N90 G1 X41.148 Z-0.01         N146 X32.846 Z-1.973          N209 X7.446
F1000.                        N147 G0 Z1.427                N210 G1 X4.572 Z-0.01
N91 Z-42.571                  N148 X27.766                  F1000.
N92 G3 X41.987 Z-43.535       N149 G1 X24.892 Z-0.01        N211 Z-1.93
R1.816                        F1000.                        N212 X7.62
N93 G1 X44.196 Z-52.941       N150 Z-1.93                   N213 X10.494 Z-0.494
N94 X47.07 Z-51.504           N151 X26.188                  N214 G0 Z1.427
N95 G0 Z1.427                 N152 G3 X27.94 Z-2.156        N215 X5.414
N96 X41.99                    R1.816                        N216 G1 X2.54 Z-0.01 F1000.
N97 G1 X39.116 Z-0.01         N153 G1 X30.814 Z-0.719       N217 Z-1.93
F1000.                        N154 G0 Z1.427                N218 X5.588
N98 Z-41.968                  N155 X25.734                  N219 X8.462 Z-0.494
N99 G3 X41.987 Z-43.535       N156 G1 X22.86 Z-0.01         N220 G0 Z1.427
R1.816                        F1000.                        N221 X4.144
N100 G1 X42.164 Z-44.289      N157 Z-1.93                   N222 G1 X1.27 Z-0.01 F1000.
N101 X45.038 Z-42.852         N158 X25.908                  N223 Z-1.93
N102 G0 Z1.427                N159 X28.782 Z-0.494          N224 X3.556
N103 X39.958                  N160 G0 Z1.427                N225 X6.43 Z-0.494
N104 G1 X37.084 Z-0.01        N161 X23.702                  N226 G0 Z1.427
F1000.                        N162 G1 X20.828 Z-0.01        N227 X2.874
N105 Z-41.93                  F1000.                        N228 G1 X0. Z-0.01 F1000.
N106 X38.38                   N163 Z-1.93                   N229 Z-1.93
N107 G3 X40.132 Z-42.156      N164 X23.876                  N230 X2.286
R1.816                        N165 X26.75 Z-0.494           N231 X5.16 Z-0.494
N108 G1 X43.006 Z-40.719      N166 G0 Z1.427                N232 G0 Z-1.51
N109 G0 Z1.427                N167 X21.67                   N233 G1 X1.274 F1000.
N110 X37.926                  N168 G1 X18.796 Z-0.01        N234 X-1.6 Z-2.946
N111 G1 X35.052 Z-0.01        F1000.                        N235 X26.188
F1000.                        N169 Z-1.93                   N236 G3 X27.584 Z-3.356
N112 Z-41.93                  N170 X21.844                  R0.8
N113 X38.1                    N171 X24.718 Z-0.494          N237 G1 X29.82 Z-5.356
N114 X40.974 Z-40.494         N172 G0 Z1.427                N238 G3 X30.023 Z-5.746
N115 G0 Z1.427                N173 X19.638                  R0.8
N116 X35.894                  N174 G1 X16.764 Z-0.01        N239 G1 Z-39.122
N117 G1 X33.02 Z-0.01         F1000.                        N240 G3 X30.017 Z-39.192
F1000.                        N175 Z-1.93                   R0.8
N118 Z-41.93                  N176 X19.812                  N241 G1 X29.36 Z-42.946
N119 X36.068                  N177 X22.686 Z-0.494          N242 X38.38
N120 X38.942 Z-40.494         N178 G0 Z1.427                N243 G3 X39.969 Z-43.653
N121 G0 Z1.427                N179 X17.606                  R0.8
N122 X33.862                  N180 G1 X14.732 Z-0.01        N244 G1 X50.369 Z-87.935
                              F1000.                        N245 X52.4
```

168

```
N246 G3 X54. Z-88.735 R0.8        N250 G0 X83.82              N253 G53 X0.
N247 G1 Z-99.454                  N251 Z5.                    N254 G53 Z0.
N248 X56.874 Z-98.017                                         N255 M30
N249 X58.064                      N252 M9                     %
```

Setup Sheet – Part 9007

Job

WCS: #0

STOCK:

DX: 2.5in
DY: 2.5in
DZ: 8.5in

PART:

DX: 2.126in
DY: 2.126in
DZ: 8.268in

STOCK LOWER IN WCS #0:
X: -1.25in

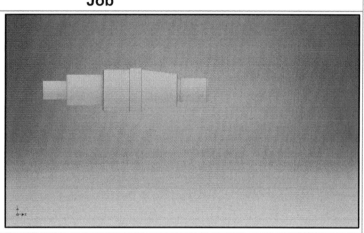

Total

NUMBER OF OPERATIONS: 1
NUMBER OF TOOLS: 1
TOOLS: T1

MAXIMUM Z: 0.197in

MINIMUM Z: -3.915in

MAXIMUM FEEDRATE: 39.37in/min
MAXIMUM SPINDLE SPEED: 500rpm

Operation 1/1

DESCRIPTION: Profile1
STRATEGY: Unspecified
WCS: #0

TOLERANCE: 0in

MAXIMUM Z: 0.197in

MINIMUM Z: -3.915in

MAXIMUM SPINDLE SPEED: 500rpm
MAXIMUM FEEDRATE: 39.37in/min
CUTTING DISTANCE: 57.532in
RAPID DISTANCE: 54.155in

T1 D0 L0

TYPE: general turning

DIAMETER: 0in

Generated by Inventor HSM Pro 4.0.0.032

169

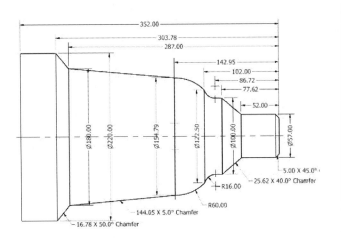
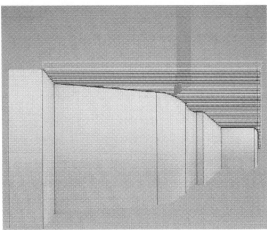

Figure 5.13 Drawing and tool path for part 0009

```
%
O0009 (Part 0009)
N10 G98 G18
N11 G21
N12 G50 S6000
N13 G28 U0.

(FACE3)
N14 T0101
N15 G54
N16 M8
N17 G98
N18 G97 S500 M3
N19 G0 X240. Z5.
N20 G0 Z-2.586
N21 G1 X222.828 F1000.
N22 X220. Z-4.
N23 X-1.6
N24 X1.228 Z-2.586
N25 G0 X240.
N26 Z5.
N27 G28 U0.

(PROFILE2)
N28 M1
N29 T0202
N30 G54
N31 G98
N32 G97 S500 M3
N33 G0 X240. Z5.
N34 G0 Z1.404
N35 X221.821
N36 G1 X220.828 F1000.
N37 X218. Z-0.01
N38 Z-306.064
N39 X219.98 Z-306.895
N40 X222.808 Z-305.481
N41 G0 Z1.404
N42 X218.828
N43 G1 X216. Z-0.01
F1000.
N44 Z-305.225
N45 X219. Z-306.484
```

```
N46 X221.828 Z-305.07
N47 G0 Z1.404
N48 X216.828
N49 G1 X214. Z-0.01
F1000.
N50 Z-304.386
N51 X217. Z-305.645
N52 X219.828 Z-304.231
N53 G0 Z1.404
N54 X214.828
N55 G1 X212. Z-0.01
F1000.
N56 Z-303.547
N57 X215. Z-304.806
N58 X217.828 Z-303.392
N59 G0 Z1.404
N60 X212.828
N61 G1 X210. Z-0.01
F1000.
N62 Z-302.708
N63 X213. Z-303.967
N64 X215.828 Z-302.552
N65 G0 Z1.404
N66 X210.828
N67 G1 X208. Z-0.01
F1000.
N68 Z-301.869
N69 X211. Z-303.128
N70 X213.828 Z-301.713
N71 G0 Z1.404
N72 X208.828
N73 G1 X206. Z-0.01
F1000.
N74 Z-301.03
N75 X209. Z-302.288
N76 X211.828 Z-300.874
N77 G0 Z1.404
N78 X206.828
N79 G1 X204. Z-0.01
F1000.
N80 Z-300.191
N81 X207. Z-301.449
N82 X209.828 Z-300.035
```

```
N83 G0 Z1.404
N84 X204.828
N85 G1 X202. Z-0.01
F1000.
N86 Z-299.352
N87 X205. Z-300.61
N88 X207.828 Z-299.196
N89 G0 Z1.404
N90 X202.828
N91 G1 X200. Z-0.01
F1000.
N92 Z-298.513
N93 X203. Z-299.771
N94 X205.828 Z-298.357
N95 G0 Z1.404
N96 X200.828
N97 G1 X198. Z-0.01
F1000.
N98 Z-297.673
N99 X201. Z-298.932
N100 X203.828 Z-297.518
N101 G0 Z1.404
N102 X198.828
N103 G1 X196. Z-0.01
F1000.
N104 Z-296.834
N105 X199. Z-298.093
N106 X201.828 Z-296.679
N107 G0 Z1.404
N108 X196.828
N109 G1 X194. Z-0.01
F1000.
N110 Z-295.995
N111 X197. Z-297.254
N112 X199.828 Z-295.84
N113 G0 Z1.404
N114 X194.828
N115 G1 X192. Z-0.01
F1000.
N116 Z-295.156
N117 X195. Z-296.415
N118 X197.828 Z-295.001
N119 G0 Z1.404
```

```
N120 X192.828
N121 G1 X190. Z-0.01
F1000.
N122 Z-294.317
N123 X193. Z-295.576
N124 X195.828 Z-294.161
N125 G0 Z1.404
N126 X190.828
N127 G1 X188. Z-0.01
F1000.
N128 Z-293.478
N129 X191. Z-294.737
N130 X193.828 Z-293.322
N131 G0 Z1.404
N132 X188.828
N133 G1 X186. Z-0.01
F1000.
N134 Z-292.639
N135 X189. Z-293.897
N136 X191.828 Z-292.483
N137 G0 Z1.404
N138 X186.828
N139 G1 X184. Z-0.01
F1000.
N140 Z-291.8
N141 X187. Z-293.058
N142 X189.828 Z-291.644
N143 G0 Z1.404
N144 X184.828
N145 G1 X182. Z-0.01
F1000.
N146 Z-290.961
N147 X185. Z-292.219
N148 X187.828 Z-290.805
N149 G0 Z1.404
N150 X182.828
N151 G1 X180. Z-0.01
F1000.
N152 Z-280.291
N153 X181.856 Z-290.9
N154 X183. Z-291.38
N155 X185.828 Z-289.966
N156 G0 Z1.404
N157 X180.828
N158 G1 X178. Z-0.01
F1000.
N159 Z-268.861
N160 X181. Z-286.006
N161 X183.828 Z-284.592
N162 G0 Z1.404
N163 X178.828
N164 G1 X176. Z-0.01
F1000.
N165 Z-257.431
N166 X179. Z-274.576
N167 X181.828 Z-273.162
N168 G0 Z1.404
N169 X176.828
N170 G1 X174. Z-0.01
F1000.
N171 Z-246.001
N172 X177. Z-263.146
N173 X179.828 Z-261.732
N174 G0 Z1.404
N175 X174.828
N176 G1 X172. Z-0.01
F1000.
N177 Z-234.571
N178 X175. Z-251.716
N179 X177.828 Z-250.302
N180 G0 Z1.404
N181 X172.828

N182 G1 X170. Z-0.01
F1000.
N183 Z-223.141
N184 X173. Z-240.286
N185 X175.828 Z-238.872
N186 G0 Z1.404
N187 X170.828
N188 G1 X168. Z-0.01
F1000.
N189 Z-211.71
N190 X171. Z-228.856
N191 X173.828 Z-227.441
N192 G0 Z1.404
N193 X168.828
N194 G1 X166. Z-0.01
F1000.
N195 Z-200.28
N196 X169. Z-217.426
N197 X171.828 Z-216.011
N198 G0 Z1.404
N199 X166.828
N200 G1 X164. Z-0.01
F1000.
N201 Z-188.85
N202 X167. Z-205.995
N203 X169.828 Z-204.581
N204 G0 Z1.404
N205 X164.828
N206 G1 X162. Z-0.01
F1000.
N207 Z-177.42
N208 X165. Z-194.565
N209 X167.828 Z-193.151
N210 G0 Z1.404
N211 X162.828
N212 G1 X160. Z-0.01
F1000.
N213 Z-165.99
N214 X163. Z-183.135
N215 X165.828 Z-181.721
N216 G0 Z1.404
N217 X160.828
N218 G1 X158. Z-0.01
F1000.
N219 Z-154.56
N220 X161. Z-171.705
N221 X163.828 Z-170.291
N222 G0 Z1.404
N223 X158.828
N224 G1 X156. Z-0.01
F1000.
N225 Z-140.753
N226 G18 G3 X156.795 Z-
147.671 R61.8
N227 G1 X159. Z-160.275
N228 X161.828 Z-158.861
N229 G0 Z1.404
N230 X156.828
N231 G1 X154. Z-0.01
F1000.
N232 Z-134.683
N233 G3 X156.795 Z-
147.671 R61.8
N234 G1 X157. Z-148.845
N235 X159.828 Z-147.431
N236 G0 Z1.404
N237 X154.828
N238 G1 X152. Z-0.01
F1000.
N239 Z-130.704
N240 G3 X155. Z-137.257
R61.8

N241 G1 X157.828 Z-
135.843
N242 G0 Z1.404
N243 X152.828
N244 G1 X150. Z-0.01
F1000.
N245 Z-127.542
N246 G3 X153. Z-132.555
R61.8
N247 G1 X155.828 Z-
131.14
N248 G0 Z1.404
N249 X150.828
N250 G1 X148. Z-0.01
F1000.
N251 Z-124.856
N252 G3 X151. Z-129.05
R61.8
N253 G1 X153.828 Z-
127.635
N254 G0 Z1.404
N255 X148.828
N256 G1 X146. Z-0.01
F1000.
N257 Z-122.492
N258 G3 X149. Z-126.151
R61.8
N259 G1 X151.828 Z-
124.737
N260 G0 Z1.404
N261 X146.828
N262 G1 X144. Z-0.01
F1000.
N263 Z-120.368
N264 G3 X147. Z-123.64
R61.8
N265 G1 X149.828 Z-
122.225
N266 G0 Z1.404
N267 X144.828
N268 G1 X142. Z-0.01
F1000.
N269 Z-118.432
N270 G3 X145. Z-121.404
R61.8
N271 G1 X147.828 Z-
119.99
N272 G0 Z1.404
N273 X142.828
N274 G1 X140. Z-0.01
F1000.
N275 Z-116.647
N276 G3 X143. Z-119.379
R61.8
N277 G1 X145.828 Z-
117.965
N278 G0 Z1.404
N279 X140.828
N280 G1 X138. Z-0.01
F1000.
N281 Z-114.991
N282 G3 X141. Z-117.522
R61.8
N283 G1 X143.828 Z-
116.108
N284 G0 Z1.404
N285 X138.828
N286 G1 X136. Z-0.01
F1000.
N287 Z-113.443
N288 G3 X139. Z-115.804
R61.8
```

N289 G1 X141.828 Z-114.39
N290 G0 Z1.404
N291 X136.828
N292 G1 X134. Z-0.01 F1000.
N293 Z-111.99
N294 G3 X137. Z-114.204 R61.8
N295 G1 X139.828 Z-112.79
N296 G0 Z1.404
N297 X134.828
N298 G1 X132. Z-0.01 F1000.
N299 Z-110.621
N300 G3 X135. Z-112.705 R61.8
N301 G1 X137.828 Z-111.291
N302 G0 Z1.404
N303 X132.828
N304 G1 X130. Z-0.01 F1000.
N305 Z-109.326
N306 G3 X133. Z-111.295 R61.8
N307 G1 X135.828 Z-109.881
N308 G0 Z1.404
N309 X130.828
N310 G1 X128. Z-0.01 F1000.
N311 Z-108.099
N312 G3 X131. Z-109.965 R61.8
N313 G1 X133.828 Z-108.55
N314 G0 Z1.404
N315 X128.828
N316 G1 X126. Z-0.01 F1000.
N317 Z-106.933
N318 G3 X129. Z-108.705 R61.8
N319 G1 X131.828 Z-107.29
N320 G0 Z1.404
N321 X126.828
N322 G1 X124. Z-0.01 F1000.
N323 Z-105.824
N324 G3 X127. Z-107.509 R61.8
N325 G1 X129.828 Z-106.095
N326 G0 Z1.404
N327 X124.828
N328 G1 X122. Z-0.01 F1000.
N329 Z-105.086
N330 G3 X123.531 Z-105.571 R1.8
N331 X125. Z-106.372 R61.8
N332 G1 X127.828 Z-104.958
N333 G0 Z1.404
N334 X122.828
N335 G1 X120. Z-0.01 F1000.
N336 Z-104.735

N337 G2 X121.969 Z-105.081 R14.2
N338 G3 X123. Z-105.338 R1.8
N339 G1 X125.828 Z-103.924
N340 G0 Z1.404
N341 X120.828
N342 G1 X118. Z-0.01 F1000.
N343 Z-104.296
N344 G2 X121. Z-104.921 R14.2
N345 G1 X123.828 Z-103.507
N346 G0 Z1.404
N347 X118.828
N348 G1 X116. Z-0.01 F1000.
N349 Z-103.761
N350 G2 X119. Z-104.527 R14.2
N351 G1 X121.828 Z-103.113
N352 G0 Z1.404
N353 X116.828
N354 G1 X114. Z-0.01 F1000.
N355 Z-103.114
N356 G2 X117. Z-104.041 R14.2
N357 G1 X119.828 Z-102.627
N358 G0 Z1.404
N359 X114.828
N360 G1 X112. Z-0.01 F1000.
N361 Z-102.338
N362 G2 X115. Z-103.452 R14.2
N363 G1 X117.828 Z-102.038
N364 G0 Z1.404
N365 X112.828
N366 G1 X110. Z-0.01 F1000.
N367 Z-101.401
N368 G2 X113. Z-102.744 R14.2
N369 G1 X115.828 Z-101.33
N370 G0 Z1.404
N371 X110.828
N372 G1 X108. Z-0.01 F1000.
N373 Z-100.251
N374 G2 X111. Z-101.892 R14.2
N375 G1 X113.828 Z-100.478
N376 G0 Z1.404
N377 X108.828
N378 G1 X106. Z-0.01 F1000.
N379 Z-98.788
N380 G2 X109. Z-100.857 R14.2
N381 G1 X111.828 Z-99.442
N382 G0 Z1.404
N383 X106.828
N384 G1 X104. Z-0.01 F1000.

N385 Z-96.756
N386 G2 X107. Z-99.568 R14.2
N387 G1 X109.828 Z-98.154
N388 G0 Z1.404
N389 X104.828
N390 G1 X102. Z-0.01 F1000.
N391 Z-82.408
N392 G3 Z-82.423 R1.8
N393 G1 Z-91.521
N394 G2 X105. Z-97.873 R14.2
N395 G1 X107.828 Z-96.459
N396 G0 Z1.404
N397 X102.828
N398 G1 X100. Z-0.01 F1000.
N399 Z-80.576
N400 X101.158 Z-81.266
N401 G3 X102. Z-82.423 R1.8
N402 G1 X104.828 Z-81.008
N403 G0 Z1.404
N404 X100.828
N405 G1 X98. Z-0.01 F1000.
N406 Z-79.384
N407 X101. Z-81.172
N408 X103.828 Z-79.757
N409 G0 Z1.404
N410 X98.828
N411 G1 X96. Z-0.01 F1000.
N412 Z-78.192
N413 X99. Z-79.98
N414 X101.828 Z-78.566
N415 G0 Z1.404
N416 X96.828
N417 G1 X94. Z-0.01 F1000.
N418 Z-77.
N419 X97. Z-78.788
N420 X99.828 Z-77.374
N421 G0 Z1.404
N422 X94.828
N423 G1 X92. Z-0.01 F1000.
N424 Z-75.809
N425 X95. Z-77.596
N426 X97.828 Z-76.182
N427 G0 Z1.404
N428 X92.828
N429 G1 X90. Z-0.01 F1000.
N430 Z-74.617
N431 X93. Z-76.405
N432 X95.828 Z-74.99
N433 G0 Z1.404
N434 X90.828
N435 G1 X88. Z-0.01 F1000.
N436 Z-73.425
N437 X91. Z-75.213
N438 X93.828 Z-73.799
N439 G0 Z1.404
N440 X88.828
N441 G1 X86. Z-0.01 F1000.
N442 Z-72.233

```
N443 X89. Z-74.021
N444 X91.828 Z-72.607
N445 G0 Z1.404
N446 X86.828
N447 G1 X84. Z-0.01
F1000.
N448 Z-71.042
N449 X87. Z-72.829
N450 X89.828 Z-71.415
N451 G0 Z1.404
N452 X84.828
N453 G1 X82. Z-0.01
F1000.
N454 Z-69.85
N455 X85. Z-71.638
N456 X87.828 Z-70.223
N457 G0 Z1.404
N458 X82.828
N459 G1 X80. Z-0.01
F1000.
N460 Z-68.658
N461 X83. Z-70.446
N462 X85.828 Z-69.032
N463 G0 Z1.404
N464 X80.828
N465 G1 X78. Z-0.01
F1000.
N466 Z-67.466
N467 X81. Z-69.254
N468 X83.828 Z-67.84
N469 G0 Z1.404
N470 X78.828
N471 G1 X76. Z-0.01
F1000.
N472 Z-66.275
N473 X79. Z-68.062
N474 X81.828 Z-66.648
N475 G0 Z1.404
N476 X76.828
N477 G1 X74. Z-0.01
F1000.
N478 Z-65.083
N479 X77. Z-66.871
N480 X79.828 Z-65.456
N481 G0 Z1.404
N482 X74.828
N483 G1 X72. Z-0.01
F1000.
N484 Z-63.891
N485 X75. Z-65.679
N486 X77.828 Z-64.265
N487 G0 Z1.404
N488 X72.828
N489 G1 X70. Z-0.01
F1000.
N490 Z-62.699
N491 X73. Z-64.487
N492 X75.828 Z-63.073
N493 G0 Z1.404
N494 X70.828
N495 G1 X68. Z-0.01
F1000.
N496 Z-61.508
N497 X71. Z-63.295
N498 X73.828 Z-61.881
N499 G0 Z1.404
N500 X68.828
N501 G1 X66. Z-0.01
F1000.
N502 Z-60.316
N503 X69. Z-62.104
N504 X71.828 Z-60.689
N505 G0 Z1.404

N506 X66.828
N507 G1 X64. Z-0.01
F1000.
N508 Z-59.124
N509 X67. Z-60.912
N510 X69.828 Z-59.498
N511 G0 Z1.404
N512 X64.828
N513 G1 X62. Z-0.01
F1000.
N514 Z-57.932
N515 X65. Z-59.72
N516 X67.828 Z-58.306
N517 G0 Z1.404
N518 X62.828
N519 G1 X60. Z-0.01
F1000.
N520 Z-56.741
N521 X63. Z-58.528
N522 X65.828 Z-57.114
N523 G0 Z1.404
N524 X60.828
N525 G1 X58. Z-0.01
F1000.
N526 Z-8.555
N527 G3 X59. Z-9.8 R1.8
N528 G1 Z-56.145
N529 X61. Z-57.337
N530 X63.828 Z-55.922
N531 G0 Z1.404
N532 X58.828
N533 G1 X56. Z-0.01
F1000.
N534 Z-7.554
N535 X57.946 Z-8.527
N536 G3 X59. Z-9.783
R1.8
N537 G1 X61.828 Z-8.369
N538 G0 Z1.404
N539 X56.828
N540 G1 X54. Z-0.01
F1000.
N541 Z-6.554
N542 X57. Z-8.054
N543 X59.828 Z-6.64
N544 G0 Z1.404
N545 X54.828
N546 G1 X52. Z-0.01
F1000.
N547 Z-5.554
N548 X55. Z-7.054
N549 X57.828 Z-5.64
N550 G0 Z1.404
N551 X52.828
N552 G1 X50. Z-0.01
F1000.
N553 Z-4.554
N554 X53. Z-6.054
N555 X55.828 Z-4.64
N556 G0 Z1.404
N557 X50.828
N558 G1 X48. Z-0.01
F1000.
N559 Z-3.554
N560 X51. Z-5.054
N561 X53.828 Z-3.64
N562 G0 Z1.404
N563 X48.828
N564 G1 X46. Z-0.01
F1000.
N565 Z-3.025
N566 G3 X47.946 Z-3.527
R1.8

N567 G1 X49. Z-4.054
N568 X51.828 Z-2.64
N569 G0 Z1.404
N570 X46.828
N571 G1 X44. Z-0.01
F1000.
N572 Z-3.
N573 X45.4
N574 G3 X47. Z-3.188
R1.8
N575 G1 X49.828 Z-1.773
N576 G0 Z1.404
N577 X44.828
N578 G1 X42. Z-0.01
F1000.
N579 Z-3.
N580 X45.
N581 X47.828 Z-1.586
N582 G0 Z1.404
N583 X42.828
N584 G1 X40. Z-0.01
F1000.
N585 Z-3.
N586 X43.
N587 X45.828 Z-1.586
N588 G0 Z1.404
N589 X40.828
N590 G1 X38. Z-0.01
F1000.
N591 Z-3.
N592 X41.
N593 X43.828 Z-1.586
N594 G0 Z1.404
N595 X38.828
N596 G1 X36. Z-0.01
F1000.
N597 Z-3.
N598 X39.
N599 X41.828 Z-1.586
N600 G0 Z1.404
N601 X36.828
N602 G1 X34. Z-0.01
F1000.
N603 Z-3.
N604 X37.
N605 X39.828 Z-1.586
N606 G0 Z1.404
N607 X34.828
N608 G1 X32. Z-0.01
F1000.
N609 Z-3.
N610 X35.
N611 X37.828 Z-1.586
N612 G0 Z1.404
N613 X32.828
N614 G1 X30. Z-0.01
F1000.
N615 Z-3.
N616 X33.
N617 X35.828 Z-1.586
N618 G0 Z1.404
N619 X30.828
N620 G1 X28. Z-0.01
F1000.
N621 Z-3.
N622 X31.
N623 X33.828 Z-1.586
N624 G0 Z1.404
N625 X28.828
N626 G1 X26. Z-0.01
F1000.
N627 Z-3.
N628 X29.
```

```
N629 X31.828 Z-1.586          N665 X19.828 Z-1.586          N701 X7.828 Z-1.586
N630 G0 Z1.404                N666 G0 Z1.404                N702 G0 Z1.404
N631 X26.828                  N667 X14.828                  N703 X2.828
N632 G1 X24. Z-0.01           N668 G1 X12. Z-0.01           N704 G1 X0. Z-0.01
F1000.                        F1000.                        F1000.
N633 Z-3.                     N669 Z-3.                     N705 Z-3.
N634 X27.                     N670 X15.                     N706 X3.
N635 X29.828 Z-1.586          N671 X17.828 Z-1.586          N707 X5.828 Z-1.586
N636 G0 Z1.404                N672 G0 Z1.404                N708 G0 Z-2.586
N637 X24.828                  N673 X12.828                  N709 G1 X1.228 F1000.
N638 G1 X22. Z-0.01           N674 G1 X10. Z-0.01           N710 X-1.6 Z-4.
F1000.                        F1000.                        N711 X45.4
N639 Z-3.                     N675 Z-3.                     N712 G3 X46.531 Z-4.234
N640 X25.                     N676 X13.                     R0.8
N641 X27.828 Z-1.586          N677 X15.828 Z-1.586          N713 G1 X56.531 Z-9.234
N642 G0 Z1.404                N678 G0 Z1.404                N714 G3 X57. Z-9.8 R0.8
N643 X22.828                  N679 X10.828                  N715 G1 Z-56.509
N644 G1 X20. Z-0.01           N680 G1 X8. Z-0.01            N716 X99.626 Z-81.908
F1000.                        F1000.                        N717 G3 X100. Z-82.423
N645 Z-3.                     N681 Z-3.                     R0.8
N646 X23.                     N682 X11.                     N718 G1 Z-91.521
N647 X25.828 Z-1.586          N683 X13.828 Z-1.586          N719 G2 X121.375 Z-
N648 G0 Z1.404                N684 G0 Z1.404                106.036 R15.2
N649 X20.828                  N685 X8.828                   N720 G3 X122.069 Z-
N650 G1 X18. Z-0.01           N686 G1 X6. Z-0.01            106.254 R0.8
F1000.                        F1000.                        N721 X154.795 Z-147.715
N651 Z-3.                     N687 Z-3.                     R60.8
N652 X21.                     N688 X9.                      N722 G1 X179.936 Z-291.4
N653 X23.828 Z-1.586          N689 X11.828 Z-1.586          N723 X219.428 Z-307.969
N654 G0 Z1.404                N690 G0 Z1.404                N724 G3 X219.982 Z-
N655 X18.828                  N691 X6.828                   308.462 R0.8
N656 G1 X16. Z-0.01           N692 G1 X4. Z-0.01            N725 G1 X222.81 Z-
F1000.                        F1000.                        307.048
N657 Z-3.                     N693 Z-3.                     N726 X223.082
N658 X19.                     N694 X7.                      N727 G0 X240.
N659 X21.828 Z-1.586          N695 X9.828 Z-1.586           N728 Z5.
N660 G0 Z1.404                N696 G0 Z1.404
N661 X16.828                  N697 X4.828                   N729 M9
N662 G1 X14. Z-0.01           N698 G1 X2. Z-0.01            N730 G28 U0. W0.
F1000.                        F1000.                        N731 M30
N663 Z-3.                     N699 Z-3.                     %
N664 X17.                     N700 X5.
```

Setup Sheet – Part 0009

Job

WCS: #0

STOCK:

DX: 220mm
DY: 220mm
DZ: 360mm

PART:

DX: 220mm
DY: 220mm
DZ: 352mm

STOCK LOWER IN WCS #0:
X: -110mm

Total

NUMBER OF OPERATIONS: 2
NUMBER OF TOOLS: 2
TOOLS: T1 T2

MAXIMUM Z: 5mm

MINIMUM Z: -308.46mm

MAXIMUM FEEDRATE: 1000mm/min
MAXIMUM SPINDLE SPEED: 500rpm

Operation 1/2

DESCRIPTION: Face3	MAXIMUM Z: 5mm	**T1** D0 L0
STRATEGY: Unspecified	MINIMUM Z: -4mm	TYPE: general turning
WCS: #0	MAXIMUM SPINDLE SPEED: 500rpm	DIAMETER: 0mm
TOLERANCE: 0.01mm	MAXIMUM FEEDRATE: 1000mm/min	LENGTH: 0mm
	CUTTING DISTANCE: 123.39mm	FLUTES: 1

Operation 2/2

DESCRIPTION: Profile2	MAXIMUM Z: 5mm	**T2** D0 L0
STRATEGY: Unspecified	MINIMUM Z: -308.46mm	TYPE: general turning
WCS: #0	MAXIMUM SPINDLE SPEED: 500rpm	DIAMETER: 0mm
TOLERANCE: 0.01mm	MAXIMUM FEEDRATE: 1000mm/min	LENGTH: 0mm
STOCK TO LEAVE: 0mm	CUTTING DISTANCE: 7506.63mm	FLUTES: 1

Generated by Inventor HSM Pro 4.0.0.032

Part 9008

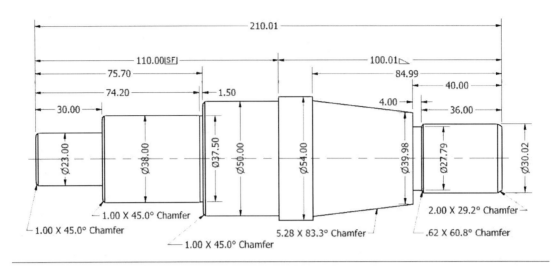

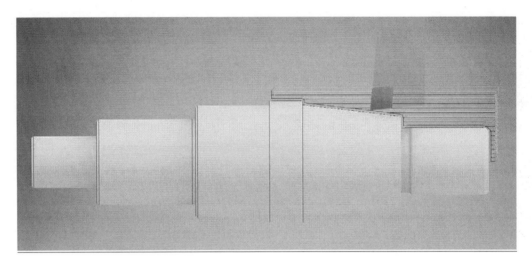

Figure 5.14 Drawing and tool path for part 9008

```
%
O9008 (Part 9008)
N10 G98 G18
N11 G21
N12 G50 S6000
N13 M31
N14 G53 G0 X0.

(Face1)
N15 T101
N16 G98
N17 M22
N18 G97 S500 M3
N19 G54
N20 M8
N21 G0 X63.5 Z5.
N22 G0 Z-1.51
```

```
N23 G1 X66.374 F1016.
N24 X63.5 Z-2.946
N25 X-1.626
N26 X1.248 Z-1.51
N27 G0 X63.5
N28 Z5.

(Profile2)
N29 G1 X63.5 Z5. F5000.
N30 G0 X64.342
N31 Z1.427
N32 G1 X61.468 Z-0.01
F1016.
N33 Z-213.767
N34 X63.48
N35 X66.354 Z-212.33
N36 G0 Z1.427
```

```
N37 X62.31
N38 G1 X59.436 Z-0.01
F1016.
N39 Z-213.767
N40 X61.468
N41 X64.342 Z-212.33
N42 G0 Z1.427
N43 X60.278
N44 G1 X57.404 Z-0.01
F1016.
N45 Z-213.767
N46 X59.436
N47 X62.31 Z-212.33
N48 G0 Z0.187
N49 G1 X58.246 Z1.427
F5000.
```

N50 X55.372 Z-0.01
F1016.
N51 Z-87.699
N52 G18 G3 X56.032 Z-88.748 R1.829
N53 G1 Z-103.767
N54 G3 X56.018 Z-103.926 R1.829
N55 G1 X55.372 Z-107.621
N56 Z-213.767
N57 X57.404
N58 X60.278 Z-212.33
N59 G0 Z0.187
N60 G1 X56.214 Z1.427 F5000.
N61 X53.34 Z-0.01 F1016.
N62 Z-86.984
N63 G3 X56.032 Z-88.748 R1.829
N64 G1 X58.906 Z-87.311
N65 G0 Z0.187
N66 G1 X54.182 Z1.427 F5000.
N67 X51.308 Z-0.01 F1016.
N68 Z-83.233
N69 X52.174 Z-86.919
N70 X52.375
N71 G3 X54.356 Z-87.21 R1.829
N72 G1 X57.23 Z-85.773
N73 G0 Z0.187
N74 G1 X52.15 Z1.427 F5000.
N75 X49.276 Z-0.01 F1016.
N76 Z-74.582
N77 X52.174 Z-86.919
N78 X52.324
N79 X55.198 Z-85.482
N80 G0 Z0.187
N81 G1 X50.118 Z1.427 F5000.
N82 X47.244 Z-0.01 F1016.
N83 Z-65.93
N84 X50.292 Z-78.907
N85 X53.166 Z-77.471
N86 G0 Z0.187
N87 G1 X48.086 Z1.427 F5000.
N88 X45.212 Z-0.01 F1016.
N89 Z-57.278
N90 X48.26 Z-70.256
N91 X51.134 Z-68.819
N92 G0 Z0.187
N93 G1 X46.054 Z1.427 F5000.
N94 X43.18 Z-0.01 F1016.
N95 Z-48.627
N96 X46.228 Z-61.604
N97 X49.102 Z-60.167
N98 G0 Z0.187
N99 G1 X44.022 Z1.427 F5000.
N100 X41.148 Z-0.01 F1016.
N101 Z-42.579
N102 G3 X41.987 Z-43.546 R1.829
N103 G1 X44.196 Z-52.953
N104 X47.07 Z-51.516

N105 G0 Z0.187
N106 G1 X41.99 Z1.427 F5000.
N107 X39.116 Z-0.01 F1016.
N108 Z-41.971
N109 G3 X41.987 Z-43.546 R1.829
N110 G1 X42.164 Z-44.301
N111 X45.038 Z-42.864
N112 G0 Z0.187
N113 G1 X39.958 Z1.427 F5000.
N114 X37.084 Z-0.01 F1016.
N115 Z-41.93
N116 X38.354
N117 G3 X40.132 Z-42.161 R1.829
N118 G1 X43.006 Z-40.724
N119 G0 Z0.187
N120 G1 X37.926 Z1.427 F5000.
N121 X35.052 Z-0.01 F1016.
N122 Z-41.93
N123 X38.1
N124 X40.974 Z-40.494
N125 G0 Z0.187
N126 G1 X35.894 Z1.427 F5000.
N127 X33.02 Z-0.01 F1016.
N128 Z-41.93
N129 X36.068
N130 X38.942 Z-40.494
N131 G0 Z0.187
N132 G1 X33.862 Z1.427 F5000.
N133 X30.988 Z-0.01 F1016.
N134 Z-4.328
N135 X31.59 Z-4.867
N136 G3 X32.055 Z-5.759 R1.829
N137 G1 Z-39.135
N138 G3 X32.041 Z-39.294 R1.829
N139 G1 X31.58 Z-41.93
N140 X34.036
N141 X36.91 Z-40.494
N142 G0 Z0.187
N143 G1 X31.83 Z1.427 F5000.
N144 X28.956 Z-0.01 F1016.
N145 Z-2.579
N146 G3 X29.355 Z-2.867 R1.829
N147 G1 X31.59 Z-4.867
N148 G3 X32.004 Z-5.455 R1.829
N149 G1 X34.878 Z-4.019
N150 G0 Z0.187
N151 G1 X29.798 Z1.427 F5000.
N152 X26.924 Z-0.01 F1016.
N153 Z-1.971
N154 G3 X29.355 Z-2.867 R1.829
N155 G1 X29.972 Z-3.419
N156 X32.846 Z-1.982

N157 G0 Z0.187
N158 G1 X27.766 Z1.427 F5000.
N159 X24.892 Z-0.01 F1016.
N160 Z-1.93
N161 X26.162
N162 G3 X27.94 Z-2.161 R1.829
N163 G1 X30.814 Z-0.724
N164 G0 Z0.187
N165 G1 X25.734 Z1.427 F5000.
N166 X22.86 Z-0.01 F1016.
N167 Z-1.93
N168 X25.908
N169 X28.782 Z-0.494
N170 G0 Z0.187
N171 G1 X23.702 Z1.427 F5000.
N172 X20.828 Z-0.01 F1016.
N173 Z-1.93
N174 X23.876
N175 X26.75 Z-0.494
N176 G0 Z0.187
N177 G1 X21.67 Z1.427 F5000.
N178 X18.796 Z-0.01 F1016.
N179 Z-1.93
N180 X21.844
N181 X24.718 Z-0.494
N182 G0 Z0.187
N183 G1 X19.638 Z1.427 F5000.
N184 X16.764 Z-0.01 F1016.
N185 Z-1.93
N186 X19.812
N187 X22.686 Z-0.494
N188 G0 Z0.187
N189 G1 X17.606 Z1.427 F5000.
N190 X14.732 Z-0.01 F1016.
N191 Z-1.93
N192 X17.78
N193 X20.654 Z-0.494
N194 G0 Z0.187
N195 G1 X15.574 Z1.427 F5000.
N196 X12.7 Z-0.01 F1016.
N197 Z-1.93
N198 X15.748
N199 X18.622 Z-0.494
N200 G0 Z0.187
N201 G1 X13.542 Z1.427 F5000.
N202 X10.668 Z-0.01 F1016.
N203 Z-1.93
N204 X13.716
N205 X16.59 Z-0.494
N206 G0 Z0.187
N207 G1 X11.51 Z1.427 F5000.
N208 X8.636 Z-0.01 F1016.
N209 Z-1.93
N210 X11.684
N211 X14.558 Z-0.494

```
N212 G0 Z0.187
N213 G1 X9.478 Z1.427
F5000.
N214 X6.604 Z-0.01
F1016.
N215 Z-1.93
N216 X9.652
N217 X12.526 Z-0.494
N218 G0 Z0.187
N219 G1 X7.446 Z1.427
F5000.
N220 X4.572 Z-0.01
F1016.
N221 Z-1.93
N222 X7.62
N223 X10.494 Z-0.494
N224 G0 Z0.187
N225 G1 X5.414 Z1.427
F5000.
N226 X2.54 Z-0.01 F1016.
N227 Z-1.93
N228 X5.588
N229 X8.462 Z-0.494
N230 G0 Z0.187
N231 G1 X4.144 Z1.427
F5000.
N232 X1.27 Z-0.01 F1016.
N233 Z-1.93
N234 X3.556
N235 X6.43 Z-0.494
N236 G0 Z0.187
N237 G1 X2.874 Z1.427
F5000.
N238 X0. Z-0.01 F1016.
N239 Z-1.93
N240 X2.286
N241 X5.16 Z-0.494
N242 X30.278
N243 G0 X58.246
N244 Z-106.184
N245 G1 X55.372 Z-
107.621 F1016.
N246 X53.34 Z-119.234
N247 Z-213.767
N248 X55.372
N249 X58.246 Z-212.33
N250 G0 Z-117.797
N251 X57.671
N252 G1 X56.214 F1016.
N253 X53.34 Z-119.234
N254 X52.032 Z-126.709
N255 Z-137.067

N256 G3 X52.018 Z-
137.226 R1.829
N257 G1 X51.308 Z-
141.284
N258 Z-213.767
N259 X53.34
N260 X56.214 Z-212.33
N261 G0 Z-139.848
N262 X55.639
N263 G1 X54.182 F1016.
N264 X51.308 Z-141.284
N265 X49.276 Z-152.897
N266 Z-213.767
N267 X51.308
N268 X54.182 Z-212.33
N269 G0 Z-151.461
N270 X53.607
N271 G1 X52.15 F1016.
N272 X49.276 Z-152.897
N273 X47.244 Z-164.51
N274 Z-213.767
N275 X49.276
N276 X52.15 Z-212.33
N277 G0 Z-163.073
N278 X51.575
N279 G1 X50.118 F1016.
N280 X47.244 Z-164.51
N281 X45.212 Z-176.123
N282 Z-213.767
N283 X47.244
N284 X50.118 Z-212.33
N285 G0 Z-174.686
N286 X49.543
N287 G1 X48.086 F1016.
N288 X45.212 Z-176.123
N289 X43.18 Z-187.736
N290 Z-213.767
N291 X45.212
N292 X48.086 Z-212.33
N293 G0 Z-186.299
N294 X47.511
N295 G1 X46.054 F1016.
N296 X43.18 Z-187.736
N297 X41.148 Z-199.349
N298 Z-213.767
N299 X43.18
N300 X46.054 Z-212.33
N301 G0 Z-197.912
N302 X45.479
N303 G1 X44.022 F1016.
N304 X41.148 Z-199.349
N305 X39.887 Z-206.558

N306 Z-213.767
N307 X41.148
N308 X44.022 Z-212.33
N309 G0 Z-205.121
N310 G1 X42.76 F1016.
N311 X39.887 Z-206.558
N312 X38.625 Z-213.767
N313 X39.887
N314 X42.76 Z-212.33
N315 G0 X56.232
N316 Z-1.51
N317 X29.644
N318 G1 X1.248 F1016.
N319 X-1.626 Z-2.946
N320 X26.162
N321 G3 X27.581 Z-3.363
R0.813
N322 G1 X29.816 Z-5.363
N323 G3 X30.023 Z-5.759
R0.813
N324 G1 Z-39.135
N325 G3 X30.017 Z-39.206
R0.813
N326 G1 X29.362 Z-42.946
N327 X38.354
N328 G3 X39.969 Z-43.664
R0.813
N329 G1 X50.366 Z-87.935
N330 X52.375
N331 G3 X54. Z-88.748
R0.813
N332 G1 Z-103.767
N333 G3 X53.994 Z-
103.838 R0.813
N334 G1 X50. Z-126.665
N335 Z-137.067
N336 G3 X49.994 Z-
137.138 R0.813
N337 G1 X36.585 Z-
213.767
N338 X39.459 Z-212.33
N339 X40.917
N340 G0 X63.5
N341 Z5.

N342 M9
N343 G53 X0.
N344 G53 Z0.
N345 M30
%
```

Setup Sheet – Part 9008

Job

WCS: #0

STOCK:

  DX: 60mm
DY: 60mm
DZ: 220mm

PART:

  DX: 54mm
DY: 54mm

  DZ: 210.01mm

STOCK LOWER IN WCS #0:

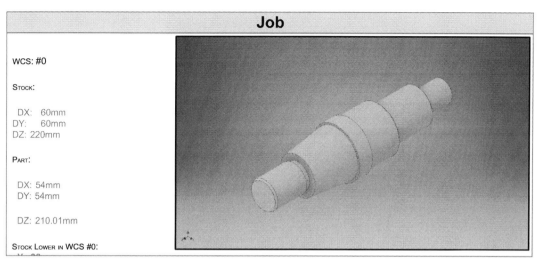

Total

NUMBER OF OPERATIONS: 2 NUMBER OF TOOLS: 1
TOOLS: T1

MAXIMUM Z: 5mm

MINIMUM Z: -213.77mm

MAXIMUM FEEDRATE: 1016mm/min MAXIMUM SPINDLE SPEED:
500rpm CUTTING DISTANCE: 2615.81mm RAPID DISTANCE:
2667.57mm ESTIMATED CYCLE TIME: 3m:21s

Operation 1/2

DESCRIPTION: Face1	MAXIMUM Z: 5mm	**T1** D0 L0
STRATEGY: Unspecified	MINIMUM Z: -2.95mm	TYPE: general turning
WCS: #0	MAXIMUM SPINDLE SPEED: 500rpm	DIAMETER: 0mm
TOLERANCE: 0.01mm	MAXIMUM FEEDRATE: 1016mm/min	LENGTH: 0mm
	CUTTING DISTANCE: 38.06mm	FLUTES: 1

Operation 2/2

DESCRIPTION: Profile2	MAXIMUM Z: 5mm	**T1** D0 L0
STRATEGY: Unspecified	MINIMUM Z: -213.77mm	TYPE: general turning
WCS: #0	MAXIMUM SPINDLE SPEED: 500rpm	DIAMETER: 0mm
TOLERANCE: 0.01mm	MAXIMUM FEEDRATE: 1016mm/min	LENGTH: 0mm
STOCK TO LEAVE: 0mm	CUTTING DISTANCE: 2577.75mm	FLUTES: 1

Generated by Inventor HSM Pro 4.0.0.032

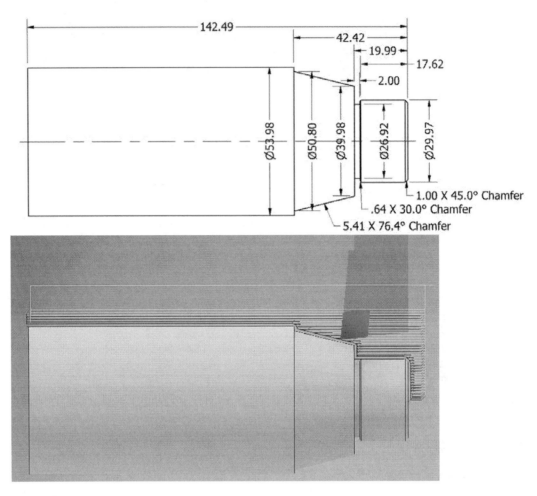

Figure 5.15 Drawing and tool path for part 0019

%
O0019 (Part 0019)
N10 G98 G18
N11 G21
N12 G50 S6000
N13 M31
N14 G53 G0 X0.

(Face7)
N15 T101
N16 G98
N17 M22
N18 G97 S500 M3
N19 G54
N20 M8
N21 G0 X83.82 Z5.
N22 G0 Z-3.516
N23 G1 X66.374 F1016.
N24 X63.5 Z-4.953
N25 X-1.626
N26 X1.248 Z-3.516
N27 G0 X83.82
N28 Z5.

(Profile3)
N29 G1 X83.82 Z5. F5000.
N30 G0 Z1.427
N31 X65.35
N32 G1 X64.342 F1016.
N33 X61.468 Z-0.01
N34 Z-148.26
N35 X63.48
N36 X66.354 Z-146.823
N37 G0 Z1.427
N38 X62.31
N39 G1 X59.436 Z-0.01
F1016.
N40 Z-148.26
N41 X61.468
N42 X64.342 Z-146.823
N43 G0 Z1.427
N44 X60.278
N45 G1 X57.404 Z-0.01
F1016.
N46 Z-148.26
N47 X59.436

N48 X62.31 Z-146.823
N49 G0 Z1.427
N50 X58.246
N51 G1 X55.372 Z-0.01
F1016.
N52 Z-47.154
N53 G18 G3 X56.007 Z-48.184 R1.829
N54 G1 Z-148.26
N55 X57.404
N56 X60.278 Z-146.823
N57 G0 Z1.427
N58 X56.214
N59 G1 X53.34 Z-0.01
F1016.
N60 Z-46.423
N61 G3 X56.007 Z-48.184
R1.829
N62 G1 X58.881 Z-46.747
N63 G0 Z1.427
N64 X54.182
N65 G1 X51.308 Z-0.01
F1016.

```
N66 Z-44.807
N67 X52.055 Z-46.355
N68 X52.349
N69 G3 X54.356 Z-46.655
R1.829
N70 G1 X57.23 Z-45.218
N71 G0 Z1.427
N72 X52.15
N73 G1 X49.276 Z-0.01
F1016.
N74 Z-40.596
N75 X52.055 Z-46.355
N76 X52.324
N77 X55.198 Z-44.918
N78 G0 Z1.427
N79 X50.118
N80 G1 X47.244 Z-0.01
F1016.
N81 Z-36.384
N82 X50.292 Z-42.701
N83 X53.166 Z-41.265
N84 G0 Z1.427
N85 X48.086
N86 G1 X45.212 Z-0.01
F1016.
N87 Z-32.172
N88 X48.26 Z-38.49
N89 X51.134 Z-37.053
N90 G0 Z1.427
N91 X46.054
N92 G1 X43.18 Z-0.01
F1016.
N93 Z-27.96
N94 X46.228 Z-34.278
N95 X49.102 Z-32.841
N96 G0 Z1.427
N97 X44.022
N98 G1 X41.148 Z-0.01
F1016.
N99 Z-24.575
N100 G3 X41.91 Z-25.327
R1.829
N101 G1 X44.196 Z-30.066
N102 X47.07 Z-28.629
N103 G0 Z1.427
N104 X41.99
N105 G1 X39.116 Z-0.01
F1016.
N106 Z-23.967
N107 G3 X41.91 Z-25.327
R1.829
N108 G1 X42.164 Z-25.854
N109 X45.038 Z-24.417
N110 G0 Z1.427
N111 X39.958
N112 G1 X37.084 Z-0.01
F1016.
N113 Z-23.927
N114 X38.354
N115 G3 X40.132 Z-24.157
R1.829
N116 G1 X43.006 Z-22.721
N117 G0 Z1.427
N118 X37.926
N119 G1 X35.052 Z-0.01
F1016.
N120 Z-23.927
N121 X38.1
N122 X40.974 Z-22.49
N123 G0 Z1.427
N124 X35.894
N125 G1 X33.02 Z-0.01
F1016.
```

```
N126 Z-23.927
N127 X36.068
N128 X38.942 Z-22.49
N129 G0 Z1.427
N130 X33.862
N131 G1 X30.988 Z-0.01
F1016.
N132 Z-5.501
N133 G3 X32.004 Z-6.766
R1.829
N134 G1 Z-23.39
N135 G3 X31.99 Z-23.549
R1.829
N136 G1 X31.924 Z-23.927
N137 X34.036
N138 X36.91 Z-22.49
N139 G0 Z1.427
N140 X31.83
N141 G1 X28.956 Z-0.01
F1016.
N142 Z-4.484
N143 X30.933 Z-5.473
N144 G3 X32.004 Z-6.758
R1.829
N145 G1 X34.878 Z-5.321
N146 G0 Z1.427
N147 X29.798
N148 G1 X26.924 Z-0.01
F1016.
N149 Z-3.96
N150 G3 X28.933 Z-4.473
R1.829
N151 G1 X29.972 Z-4.992
N152 X32.846 Z-3.555
N153 G0 Z1.427
N154 X27.766
N155 G1 X24.892 Z-0.01
F1016.
N156 Z-3.937
N157 X26.346
N158 G3 X27.94 Z-4.12
R1.829
N159 G1 X30.814 Z-2.683
N160 G0 Z1.427
N161 X25.734
N162 G1 X22.86 Z-0.01
F1016.
N163 Z-3.937
N164 X25.908
N165 X28.782 Z-2.5
N166 G0 Z1.427
N167 X23.702
N168 G1 X20.828 Z-0.01
F1016.
N169 Z-3.937
N170 X23.876
N171 X26.75 Z-2.5
N172 G0 Z1.427
N173 X21.67
N174 G1 X18.796 Z-0.01
F1016.
N175 Z-3.937
N176 X21.844
N177 X24.718 Z-2.5
N178 G0 Z1.427
N179 X19.638
N180 G1 X16.764 Z-0.01
F1016.
N181 Z-3.937
N182 X19.812
N183 X22.686 Z-2.5
N184 G0 Z1.427
N185 X17.606
```

```
N186 G1 X14.732 Z-0.01
F1016.
N187 Z-3.937
N188 X17.78
N189 X20.654 Z-2.5
N190 G0 Z1.427
N191 X15.574
N192 G1 X12.7 Z-0.01
F1016.
N193 Z-3.937
N194 X15.748
N195 X18.622 Z-2.5
N196 G0 Z1.427
N197 X13.542
N198 G1 X10.668 Z-0.01
F1016.
N199 Z-3.937
N200 X13.716
N201 X16.59 Z-2.5
N202 G0 Z1.427
N203 X11.51
N204 G1 X8.636 Z-0.01
F1016.
N205 Z-3.937
N206 X11.684
N207 X14.558 Z-2.5
N208 G0 Z1.427
N209 X9.478
N210 G1 X6.604 Z-0.01
F1016.
N211 Z-3.937
N212 X9.652
N213 X12.526 Z-2.5
N214 G0 Z1.427
N215 X7.446
N216 G1 X4.572 Z-0.01
F1016.
N217 Z-3.937
N218 X7.62
N219 X10.494 Z-2.5
N220 G0 Z1.427
N221 X5.414
N222 G1 X2.54 Z-0.01
F1016.
N223 Z-3.937
N224 X5.588
N225 X8.462 Z-2.5
N226 G0 Z1.427
N227 X4.144
N228 G1 X1.27 Z-0.01
F1016.
N229 Z-3.937
N230 X3.556
N231 X6.43 Z-2.5
N232 G0 Z1.427
N233 X2.874
N234 G1 X0. Z-0.01 F1016.
N235 Z-3.937
N236 X2.286
N237 X5.16 Z-2.5
N238 G0 Z-3.516
N239 G1 X1.248 F1016.
N240 X-1.626 Z-4.953
N241 X26.346
N242 G3 X27.496 Z-5.191
R0.813
N243 G1 X29.496 Z-6.191
N244 G3 X29.972 Z-6.766
R0.813
N245 G1 Z-23.39
N246 G3 X29.966 Z-23.461
R0.813
N247 G1 X29.706 Z-24.943
```

```
N248 X38.354              N253 G1 Z-148.26           N259 G53 X0.
N249 G3 X39.934 Z-25.565  N254 X56.849 Z-146.823     N260 G53 Z0.
R0.813                    N255 X58.039               N261 M30
N250 G1 X50.454 Z-47.371  N256 G0 X83.82             %
N251 X52.349              N257 Z5.
N252 G3 X53.975 Z-48.184
R0.813                    N258 M9
```

Setup Sheet – Part 0019

Job

WCS: #0

STOCK:

DX: 60mm
DY: 60mm
DZ: 150mm

PART:

DX: 53.98mm
DY: 53.98mm
DZ: 142.49mm

STOCK LOWER IN WCS #0:
X: -30mm

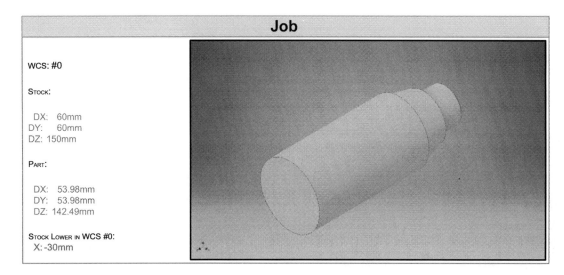

Total

NUMBER OF OPERATIONS: 2
NUMBER OF TOOLS: 1
TOOLS: T1

MAXIMUM Z: 5mm

MINIMUM Z: -148.26mm

MAXIMUM FEEDRATE: 1016mm/min
MAXIMUM SPINDLE SPEED: 500rpm

Operation 1/2

DESCRIPTION: Face7   MAXIMUM Z: 5mm   T1 D0 L0

STRATEGY: Unspecified   MINIMUM Z: -4.95mm   TYPE: general turning
WCS: #0   MAXIMUM SPINDLE SPEED: 500rpm   DIAMETER: 0mm
TOLERANCE: 0.01mm   MAXIMUM FEEDRATE: 1016mm/min   LENGTH: 0mm

                             CUTTING DISTANCE: 45.35mm   FLUTES: 1

Operation 2/2

DESCRIPTION: Profile3   MAXIMUM Z: 5mm   T1 D0 L0

STRATEGY: Unspecified   MINIMUM Z: -148.26mm   TYPE: general turning
WCS: #0   MAXIMUM SPINDLE SPEED: 500rpm   DIAMETER: 0mm
TOLERANCE: 0.01mm   MAXIMUM FEEDRATE: 1016mm/min   LENGTH: 0mm

STOCK TO LEAVE: 0mm   CUTTING DISTANCE: 1405mm   FLUTES: 1

Generated by Inventor HSM Pro 4.0.0.032

182
```
```

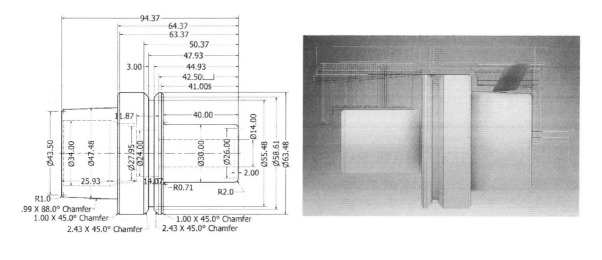

Figure 5.16 Drawing and tool path for part 1001

```
%
O1001 (Part 1001)
N10 G94
N15 G26 S6000
N20 G14
N25 M09
N30 T101
N35 M09
N40 G94
N45 G97 S500 M03
N50 G00 X90. Z5.
N55 G00 Z-1.403
N60 G01 X72.828 F1000.
N65 X70. Z-2.817
N70 X-1.6
N75 X1.228 Z-1.403
N80 G00 X90.
N85 Z5.
N90 G14
N95 M09
N100 M01
N105 T808
N110 M08
N115 G94
N120 G97 S500 M03
N125 G00 X71.657 Z3.002
N130 G01 X66. Z0.173
F1000.
N135 Z-30.942
N140 X67.476 Z-31.68
N145 X67.728 Z-31.812
N150 X67.966 Z-31.95
N155 X68.191 Z-32.093
N160 X68.483 Z-32.301
N165 X68.749 Z-32.517
N170 X68.989 Z-32.741
N175 X69.202 Z-32.972
N180 X69.387 Z-33.208
N185 X69.544 Z-33.45
N190 X69.671 Z-33.696
N195 X69.768 Z-33.945
N200 X69.836 Z-34.197
N205 X69.873 Z-34.45
N210 X69.881 Z-34.617
```

```
N215 Z-100.983
N220 X69.98
N225 X75.637 Z-98.155
N230 G00 Z3.002
N235 X67.657
N240 G01 X62. Z0.173
F1000.
N245 Z-29.665
N250 X62.477 Z-29.741
N255 X62.943 Z-29.831
N260 X63.398 Z-29.935
N265 X63.838 Z-30.053
N270 X64.263 Z-30.185
N275 X64.671 Z-30.33
N280 X65.06 Z-30.487
N285 X65.272 Z-30.582
N290 X65.476 Z-30.68
N295 X67.476 Z-31.68
N300 X67.746 Z-31.822
N305 X68. Z-31.97
N310 X73.657 Z-29.142
N315 G00 Z3.002
N320 X63.657
N325 G01 X58. Z0.173
F1000.
N330 Z-29.517
N335 X59.881
N340 X60.24 Z-29.521
N345 X60.599 Z-29.534
N350 X60.957 Z-29.555
N355 X61.415 Z-29.595
N360 X61.868 Z-29.647
N365 X62.314 Z-29.713
N370 X62.752 Z-29.792
N375 X63.179 Z-29.883
N380 X63.596 Z-29.986
N385 X64. Z-30.101
N390 X69.657 Z-27.273
N395 G00 Z0.383
N400 X59.657 Z3.002
N405 G01 X54. Z0.173
F1000.
N410 Z-29.517
N415 X59.881
```

```
N420 X60.
N425 X65.657 Z-26.689
N430 G00 Z0.383
N435 X55.657 Z3.002
N440 G01 X50. Z0.173
F1000.
N445 Z-1.622
N450 X50.307 Z-1.852
N455 X50.589 Z-2.089
N460 X50.846 Z-2.333
N465 X51.075 Z-2.584
N470 X51.278 Z-2.841
N475 X51.452 Z-3.102
N480 X51.598 Z-3.369
N485 X51.716 Z-3.638
N490 X51.804 Z-3.911
N495 X51.863 Z-4.185
N500 X51.893 Z-4.46
N505 X53.605 Z-29.517
N510 X56.
N515 X61.657 Z-26.688
N520 G00 Z0.383
N525 X53.301 Z3.002
N530 G01 X47.644 Z0.173
F1000.
N535 Z-0.437
N540 X48.069 Z-0.598
N545 X48.476 Z-0.77
N550 X48.864 Z-0.952
N555 X48.891 Z-0.965
N560 X49.273 Z-1.167
N565 X49.633 Z-1.38
N570 X49.97 Z-1.601
N575 X50.282 Z-1.832
N580 X50.569 Z-2.07
N585 X50.829 Z-2.316
N590 X51.062 Z-2.569
N595 X51.268 Z-2.827
N600 X51.445 Z-3.091
N605 X51.594 Z-3.359
N610 X51.713 Z-3.631
N615 X51.803 Z-3.906
N620 X51.863 Z-4.183
N625 X51.893 Z-4.46
```

```
N630 X52. Z-6.031
N635 X57.657 Z-3.203
N640 G00 Z0.383
N645 X50.997 Z3.002
N650 G01 X45.34 Z0.174
F1000.
N655 X45.829 Z0.075
N660 X46.306 Z-0.036
N665 X46.771 Z-0.159
N670 X47.222 Z-0.295
N675 X47.658 Z-0.443
N680 X48.078 Z-0.602
N685 X48.481 Z-0.772
N690 X48.864 Z-0.952
N695 X48.891 Z-0.965
N700 X49.279 Z-1.171
N705 X49.644 Z-1.386
N710 X55.301 Z1.442
N715 G00 X-1.6
N720 Z-2.517
N725 G01 X41.898 F1000.
N730 X41.899
N735 X42.238 Z-2.524
N740 X42.573 Z-2.546
N745 X42.904 Z-2.583
N750 X43.227 Z-2.634
N755 X43.54 Z-2.699
N760 X43.841 Z-2.777
N765 X44.126 Z-2.868
N770 X44.395 Z-2.971
N775 X44.645 Z-3.085
N780 X44.889 Z-3.218
N785 X45.11 Z-3.361
N790 X45.306 Z-3.513
N795 X45.476 Z-3.672
N800 X45.618 Z-3.839
N805 X45.732 Z-4.01
N810 X45.816 Z-4.186
N815 X45.871 Z-4.365
N820 X45.895 Z-4.545
N825 X47.807 Z-32.517
N830 X59.881
N835 X60.095 Z-32.524
N840 X60.307 Z-32.543
N845 X60.513 Z-32.574
N850 X60.711 Z-32.615
N855 X60.898 Z-32.668
N860 X61.074 Z-32.73
N865 X61.234 Z-32.801
N870 X63.234 Z-33.801
N875 X63.4 Z-33.899
N880 X63.544 Z-34.005
N885 X63.664 Z-34.119
N890 X63.759 Z-34.238
N895 X63.827 Z-34.362
N900 X63.868 Z-34.489
N905 X63.881 Z-34.617
N910 Z-100.983
N915 X69.537 Z-98.155
N920 X69.881
N925 G00 X70.
N930 Z5.
N935 G14
N940 M09
N945 M01
N950 T606
N955 M09
N960 G94
N965 G97 S500 M03
N970 G00 X90. Z5.
N975 G00 Z-51.783
N980 X67.881
N985 G01 X63.881 F1000.
```

```
N990 X63.439
N995 X67.881
N1000 G00 Z-52.75
N1005 G01 X63.881 F1000.
N1010 X61.505
N1015 X63.863
N1020 X63.881
N1025 X67.439
N1030 G00 X67.881
N1035 Z-53.717
N1040 G01 X63.881 F1000.
N1045 X63.439
N1050 X63.881
N1055 X66.845
N1060 G00 Z-50.219
N1065 G01 X63.881 Z-51.562
F1000.
N1070 X61.505 Z-52.75
N1075 X63.881
N1080 Z-53.938
N1085 X61.505 Z-52.75
N1090 X62.543
N1095 Z-50.817
N1100 X58.677 Z-52.75
N1105 X62.543 Z-54.683
N1110 X65.372 Z-53.269
N1115 G00 X90.
N1120 Z5.
N1125 G14
N1130 M09
N1135 M01
N1140 T808
N1145 M08
N1150 G94
N1155 G97 S500 M03
N1160 G00 X90. Z-108.983
N1165 G00 X73.98
N1170 G01 X69.98 F1000.
N1175 X68.
N1180 Z-57.193
N1185 X70.828 Z-58.607
N1190 G00 X72.
N1195 Z-108.983
N1200 G01 X68. F1000.
N1205 X66.
N1210 Z-57.193
N1215 X68.828 Z-58.607
N1220 X69.225
N1225 G00 X70.
N1230 Z-108.983
N1235 G01 X66. F1000.
N1240 X64.
N1245 Z-65.523
N1250 X65.458 Z-57.193
N1255 X68.286 Z-58.607
N1260 G00 Z-108.983
N1265 X68.
N1270 G01 X64. F1000.
N1275 X62.
N1280 Z-76.954
N1285 X65. Z-59.808
N1290 X67.828 Z-61.223
N1295 G00 Z-108.983
N1300 X66.
N1305 G01 X62. F1000.
N1310 X60.
N1315 Z-88.384
N1320 X63. Z-71.239
N1325 X65.828 Z-72.653
N1330 G00 Z-108.983
N1335 X64.
N1340 G01 X60. F1000.
N1345 X58.198
```

```
N1350 Z-98.683
N1355 X61. Z-82.669
N1360 X63.828 Z-84.083
N1365 G00 Z-108.983
N1370 X62.198
N1375 G01 X58.198 F1000.
N1380 X56.396
N1385 X59.198 Z-92.968
N1390 X62.026 Z-94.382
N1395 G00 Z-110.397
N1400 X57.216
N1405 G01 X54.388 Z-
108.983 F1000.
N1410 X63.45 Z-57.193
N1415 X66.278 Z-58.607
N1420 X67.218
N1425 G00 X90.
N1430 Z5.
N1435 G14
N1440 M09
N1445 M01
N1450 T606
N1455 M09
N1460 G94
N1465 G97 S500 M03
N1470 G00 X90. Z5.
N1475 G00 Z-61.183
N1480 X74.
N1485 G01 X70. F1000.
N1490 X63.372
N1495 X74.
N1500 G00 Z-62.166
N1505 G01 X70. F1000.
N1510 X60.383
N1515 X73.484
N1520 G00 X74.
N1525 Z-63.148
N1530 G01 X70. F1000.
N1535 X32.001
N1540 X68.516
N1545 X70.
N1550 X73.482
N1555 G00 X74.
N1560 Z-64.13
N1565 G01 X70. F1000.
N1570 X32.001
N1575 X61.836
N1580 X63.836 Z-63.13
N1585 X70.906
N1590 G00 X74.
N1595 Z-65.113
N1600 G01 X70. F1000.
N1605 X32.001
N1610 X58.867
N1615 X60.867 Z-64.113
N1620 X70.525
N1625 G00 X74.
N1630 Z-66.095
N1635 G01 X70. F1000.
N1640 X32.001
N1645 X34.001 Z-65.095
N1650 X70.525
N1655 G00 X74.
N1660 Z-67.077
N1665 G01 X70. F1000.
N1670 X32.001
N1675 X34.001 Z-66.077
N1680 X70.525
N1685 G00 X74.
N1690 Z-68.06
N1695 G01 X70. F1000.
N1700 X32.001
N1705 X34.001 Z-67.06
```

```
N1710 X70.525
N1715 G00 X74.
N1720 Z-69.042
N1725 G01 X70. F1000.
N1730 X32.001
N1735 X34.001 Z-68.042
N1740 X70.525
N1745 G00 X74.
N1750 Z-70.025
N1755 G01 X70. F1000.
N1760 X32.001
N1765 X34.001 Z-69.025
N1770 X70.525
N1775 G00 X74.
N1780 Z-71.007
N1785 G01 X70. F1000.
N1790 X32.001
N1795 X34.001 Z-70.007
N1800 X70.525
N1805 G00 X74.
N1810 Z-71.989
N1815 G01 X70. F1000.
N1820 X32.001
N1825 X34.001 Z-70.989
N1830 X70.525
N1835 G00 X74.
N1840 Z-72.972
N1845 G01 X70. F1000.
N1850 X32.001
N1855 X34.001 Z-71.972
N1860 X70.525
N1865 G00 X74.
N1870 Z-73.954
N1875 G01 X70. F1000.
N1880 X32.001
N1885 X34.001 Z-72.954
N1890 X70.525
N1895 G00 X74.
N1900 Z-74.936
N1905 G01 X70. F1000.
N1910 X32.001
N1915 X34.001 Z-73.936
N1920 X70.525
N1925 G00 X74.
N1930 Z-75.919
N1935 G01 X70. F1000.
N1940 X32.001
N1945 X34.001 Z-74.919
N1950 X70.525
N1955 G00 X74.
N1960 Z-76.901
N1965 G01 X70. F1000.
N1970 X32.001
N1975 X34.001 Z-75.901
N1980 X70.525
N1985 G00 X74.
N1990 Z-77.883
N1995 G01 X70. F1000.
N2000 X32.001
N2005 X34.001 Z-76.883
N2010 X70.525
N2015 G00 X74.
N2020 Z-78.866
N2025 G01 X70. F1000.
N2030 X32.001
N2035 X34.001 Z-77.866
N2040 X70.525
N2045 G00 X74.
N2050 Z-79.848
N2055 G01 X70. F1000.
N2060 X32.001
N2065 X34.001 Z-78.848
N2070 X70.525
N2075 G00 X74.
N2080 Z-80.831
N2085 G01 X70. F1000.
N2090 X32.001
N2095 X34.001 Z-79.831
N2100 X70.525
N2105 G00 X74.
N2110 Z-81.813
N2115 G01 X70. F1000.
N2120 X32.001
N2125 X34.001 Z-80.813
N2130 X70.525
N2135 G00 X74.
N2140 Z-82.795
N2145 G01 X70. F1000.
N2150 X32.001
N2155 X34.001 Z-81.795
N2160 X70.525
N2165 G00 X74.
N2170 Z-83.778
N2175 G01 X70. F1000.
N2180 X32.001
N2185 X34.001 Z-82.778
N2190 X70.525
N2195 G00 X74.
N2200 Z-84.76
N2205 G01 X70. F1000.
N2210 X32.001
N2215 X34.001 Z-83.76
N2220 X70.525
N2225 G00 X74.
N2230 Z-85.742
N2235 G01 X70. F1000.
N2240 X32.001
N2245 X34.001 Z-84.742
N2250 X70.525
N2255 G00 X74.
N2260 Z-86.725
N2265 G01 X70. F1000.
N2270 X32.001
N2275 X34.001 Z-85.725
N2280 X70.525
N2285 G00 X74.
N2290 Z-87.707
N2295 G01 X70. F1000.
N2300 X32.001
N2305 X34.001 Z-86.707
N2310 X70.525
N2315 G00 X74.
N2320 Z-88.69
N2325 G01 X70. F1000.
N2330 X32.001
N2335 X34.001 Z-87.69
N2340 X70.525
N2345 G00 X74.
N2350 Z-89.672
N2355 G01 X70. F1000.
N2360 X32.001
N2365 X34.001 Z-88.672
N2370 X70.525
N2375 G00 X74.
N2380 Z-90.654
N2385 G01 X70. F1000.
N2390 X32.001
N2395 X34.001 Z-89.654
N2400 X70.525
N2405 G00 X74.
N2410 Z-91.637
N2415 G01 X70. F1000.
N2420 X32.001
N2425 X34.001 Z-90.637
N2430 X70.525
N2435 G00 X74.
N2440 Z-92.619
N2445 G01 X70. F1000.
N2450 X32.001
N2455 X34.001 Z-91.619
N2460 X70.525
N2465 G00 X74.
N2470 Z-93.601
N2475 G01 X70. F1000.
N2480 X32.001
N2485 X34.001 Z-92.601
N2490 X70.525
N2495 G00 X74.
N2500 Z-94.584
N2505 G01 X70. F1000.
N2510 X32.001
N2515 X34.001 Z-93.584
N2520 X70.525
N2525 G00 X74.
N2530 Z-95.566
N2535 G01 X70. F1000.
N2540 X32.001
N2545 X34.001 Z-94.566
N2550 X70.525
N2555 G00 X74.
N2560 Z-96.549
N2565 G01 X70. F1000.
N2570 X32.001
N2575 X34.001 Z-95.549
N2580 X70.525
N2585 G00 X74.
N2590 Z-97.531
N2595 G01 X70. F1000.
N2600 X32.001
N2605 X34.001 Z-96.531
N2610 X70.512
N2615 G00 X74.
N2620 Z-98.513
N2625 G01 X70. F1000.
N2630 X31.996
N2635 X32.145
N2640 X34.145 Z-97.513
N2645 X70.512
N2650 G00 X74.
N2655 Z-99.496
N2660 G01 X70. F1000.
N2665 X31.668
N2670 X32.117
N2675 X34.117 Z-98.496
N2680 X70.512
N2685 G00 X74.
N2690 Z-100.478
N2695 G01 X70. F1000.
N2700 X30.742
N2705 X32.122
N2710 X34.122 Z-99.478
N2715 X70.512
N2720 G00 X74.
N2725 Z-101.46
N2730 G01 X70. F1000.
N2735 X28.86
N2740 X31.752
N2745 X33.752 Z-100.46
N2750 X70.512
N2755 G00 X74.
N2760 Z-102.443
N2765 G01 X70. F1000.
N2770 X16.003
N2775 X30.767
N2780 X32.767 Z-101.443
N2785 X70.525
N2790 G00 X74.
N2795 Z-103.425
N2800 G01 X70. F1000.
```

```
N2805 X16.003
N2810 X29.291
N2815 X31.291 Z-102.425
N2820 X70.525
N2825 G00 X74.
N2830 Z-104.407
N2835 G01 X70. F1000.
N2840 X16.003
N2845 X27.393
N2850 X29.393 Z-103.407
N2855 X70.525
N2860 G00 X74.
N2865 Z-105.39
N2870 G01 X70. F1000.
N2875 X16.003
N2880 X18.003 Z-104.39
N2885 X70.525
N2890 G00 X74.
N2895 Z-106.372
N2900 G01 X70. F1000.
N2905 X16.003
N2910 X18.003 Z-105.372
N2915 X70.525
N2920 G00 X74.
N2925 Z-107.355
N2930 G01 X70. F1000.
N2935 X16.003
N2940 X18.003 Z-106.355
N2945 X67.374
N2950 G00 Z-61.183
N2955 G01 X67.372 F1000.
N2960 X63.372
N2965 X62.426 Z-61.656
N2970 X62.195 Z-61.762
N2975 X61.945 Z-61.858
N2980 X61.681 Z-61.942
N2985 X61.402 Z-62.015
N2990 X61.112 Z-62.075
N2995 X60.812 Z-62.122
N3000 X60.506 Z-62.156
N3005 X60.194 Z-62.176
N3010 X59.881 Z-62.183
N3015 X32.001
N3020 Z-98.383
N3025 X31.986 Z-98.622
N3030 X31.941 Z-98.859
N3035 X31.866 Z-99.095
N3040 X31.762 Z-99.328
N3045 X31.629 Z-99.557
N3050 X31.467 Z-99.782
N3055 X31.277 Z-100.001
N3060 X31.061 Z-100.214
N3065 X30.817 Z-100.419
N3070 X30.549 Z-100.617
N3075 X30.256 Z-100.805
N3080 X29.941 Z-100.984
N3085 X29.603 Z-101.153
N3090 X29.245 Z-101.311
N3095 X28.868 Z-101.457
N3100 X28.473 Z-101.592
N3105 X28.062 Z-101.713
N3110 X27.637 Z-101.822
N3115 X27.198 Z-101.916
N3120 X26.749 Z-101.997
N3125 X26.291 Z-102.064
N3130 X25.825 Z-102.116
N3135 X25.353 Z-102.153
N3140 X24.878 Z-102.176
N3145 X24.401 Z-102.183
N3150 X16.003
N3155 Z-107.355
N3160 X18.831 Z-105.94
N3165 X72.806

N3170 G00 X73.982
N3175 Z-61.183
N3180 X73.98
N3185 G01 X36.001 F1000.
N3190 X32.001
N3195 X30.401
N3200 X30.297 Z-61.19
N3205 X30.201 Z-61.21
N3210 X30.118 Z-61.242
N3215 X30.054 Z-61.283
N3220 X30.014 Z-61.331
N3225 X30.001 Z-61.383
N3230 Z-98.383
N3235 X29.985 Z-98.592
N3240 X29.938 Z-98.801
N3245 X29.86 Z-99.006
N3250 X29.752 Z-99.209
N3255 X29.613 Z-99.406
N3260 X29.446 Z-99.598
N3265 X29.25 Z-99.783
N3270 X29.028 Z-99.96
N3275 X28.779 Z-100.129
N3280 X28.506 Z-100.288
N3285 X28.21 Z-100.436
N3290 X27.892 Z-100.572
N3295 X27.555 Z-100.697
N3300 X27.201 Z-100.808
N3305 X26.83 Z-100.906
N3310 X26.447 Z-100.99
N3315 X26.051 Z-101.059
N3320 X25.647 Z-101.113
N3325 X25.235 Z-101.152
N3330 X24.819 Z-101.175
N3335 X24.401 Z-101.183
N3340 X10.803
N3345 X14.803
N3350 X17.602
N3355 G00 X90.
N3360 Z5.
N3365 M09
N3370 G94
N3375 G97 S500 M03
N3380 G00 X90. Z5.
N3385 Z-101.183
N3390 G01 X-1.6 F1000.
N3395 X90.
N3400 G00 Z5.
N3405 G14
N3410 M09
N3415 M01
N3420 T400
N3425 M09
N3430 G94
N3435 G97 S1000 M03
N3440 G00 X0. Z15.
N3445 G00 Z5.
N3450 Z2.183
N3455 G01 Z-102.591 F250.
N3460 Z2.183
N3465 G00 Z5.
N3470 Z15.
N3475 G14
N3480 M09
N3485 M01
N3490 T500
N3495 M09
N3500 G94
N3505 G97 S1000 M03
N3510 G00 X0. Z15.
N3515 G00 Z5.
N3520 X-22.
N3525 Z-49.683
N3530 G01 Z-54.683 F0.

N3535 G13 X0. Z11. R11.
N3540 Z-11. R11.
N3545 Z11. R11.
N3550 Z-11. R11.
N3555 Z11. R11.
N3560 Z-11. R11.
N3565 Z11. R11.
N3570 Z-11. R11.
N3575 Z11. R11.
N3580 Z-11. R11.
N3585 Z11. R11.
N3590 Z-11. R11.
N3595 Z11. R11.
N3600 Z-11. R11.
N3605 Z11. R11.
N3610 Z-11. R11.
N3615 Z11. R11.
N3620 Z-11. R11.
N3625 Z11. R11.
N3630 Z-11. R11.
N3635 Z11. R11.
N3640 Z-11. R11.
N3645 Z11. R11.
N3650 Z-11. R11.
N3655 Z11. R11.
N3660 Z-11. R11.
N3665 Z11. R11.
N3670 Z-11. R11.
N3675 Z11. R11.
N3680 Z-11. R11.
N3685 Z11. R11.
N3690 Z-11. R11.
N3695 Z11. R11.
N3700 Z-11. R11.
N3705 Z11. R11.
N3710 Z-11. R11.
N3715 Z11. R11.
N3720 Z-11. R11.
N3725 Z11. R11.
N3730 Z-11. R11.
N3735 Z11. R11.
N3740 Z-11. R11.
N3745 Z11. R11.
N3750 Z-11. R11.
N3755 Z11. R11.
N3760 Z-11. R11.
N3765 Z11. R11.
N3770 Z-11. R11.
N3775 Z11. R11.
N3780 Z-11. R11.
N3785 Z11. R11.
N3790 Z-11. R11.
N3795 Z11. R11.
N3800 Z-11. R11.
N3805 Z11. R11.
N3810 Z-11. R11.
N3815 Z11. R11.
N3820 Z-11. R11.
N3825 Z11. R11.
N3830 Z-11. R11.
N3835 Z11. R11.
N3840 Z-11. R11.
N3845 Z11. R11.
N3850 Z-11. R11.
N3855 Z11. R11.
N3860 Z-11. R11.
N3865 Z11. R11.
N3870 Z-11. R11.
N3875 Z11. R11.
N3880 Z-11. R11.
N3885 Z11. R11.
N3890 Z-11. R11.
N3895 Z11. R11.
```

```
N3900 Z-11. R11.
N3905 Z11. R11.
N3910 Z-11. R11.
N3915 Z11. R11.
N3920 Z-11. R11.
N3925 Z11. R11.
N3930 Z-11. R11.
N3935 Z11. R11.
N3940 Z-11. R11.
N3945 Z11. R11.
N3950 Z0. R5.5
N3955 G00 Z5.
N3960 Z15.
N3965 G14
N3970 M09
N3975 M01
N3980 T700
N3985 M09
N3990 G94
N3995 G97 S1000 M03
N4000 G00 X0. Z15.
N4005 G00 Z5.
N4010 X10.012
N4015 Z2.183
N4020 G01 Z-2.817 F0.
N4025 G13 X-11.17 Z-5.006 R7.5
N4030 X11.17 Z5.006 R7.5
N4035 X-11.17 Z-5.006 R7.5
N4040 X11.17 Z5.006 R7.5
N4045 X-11.17 Z-5.006 R7.5
N4050 X11.17 Z5.006 R7.5
N4055 X-11.17 Z-5.006 R7.5
N4060 X11.17 Z5.006 R7.5
N4065 X-11.17 Z-5.006 R7.5
N4070 X11.17 Z5.006 R7.5
N4075 X-11.17 Z-5.006 R7.5
N4080 X11.17 Z5.006 R7.5
N4085 X-11.17 Z-5.006 R7.5
N4090 X11.17 Z5.006 R7.5
N4095 X-11.17 Z-5.006 R7.5
N4100 X11.17 Z5.006 R7.5
N4105 X-11.17 Z-5.006 R7.5
N4110 X11.17 Z5.006 R7.5
N4115 X-11.17 Z-5.006 R7.5
N4120 X11.17 Z5.006 R7.5
N4125 X-11.17 Z-5.006 R7.5
N4130 X11.17 Z5.006 R7.5
N4135 X-11.17 Z-5.006 R7.5
N4140 X11.17 Z5.006 R7.5
N4145 X-11.17 Z-5.006 R7.5
N4150 X11.17 Z5.006 R7.5
N4155 X-11.17 Z-5.006 R7.5
N4160 X11.17 Z5.006 R7.5
N4165 X-11.17 Z-5.006 R7.5
N4170 X11.17 Z5.006 R7.5
N4175 X-11.17 Z-5.006 R7.5
N4180 X11.17 Z5.006 R7.5
N4185 X-11.17 Z-5.006 R7.5
N4190 X11.17 Z5.006 R7.5
N4195 X-11.17 Z-5.006 R7.5
N4200 X11.17 Z5.006 R7.5
N4205 X-11.17 Z-5.006 R7.5
N4210 X11.17 Z5.006 R7.5
N4215 X-11.17 Z-5.006 R7.5
N4220 X11.17 Z5.006 R7.5
N4225 X-11.17 Z-5.006 R7.5
N4230 X11.17 Z5.006 R7.5
N4235 X-11.17 Z-5.006 R7.5
N4240 X11.17 Z5.006 R7.5
N4245 X-11.17 Z-5.006 R7.5
N4250 X11.17 Z5.006 R7.5
N4255 X-11.17 Z-5.006 R7.5
N4260 X11.17 Z5.006 R7.5
N4265 X-11.17 Z-5.006 R7.5
N4270 X11.17 Z5.006 R7.5
N4275 X-11.17 Z-5.006 R7.5
N4280 X11.17 Z5.006 R7.5
N4285 X-11.17 Z-5.006 R7.5
N4290 X11.17 Z5.006 R7.5
N4295 X-11.17 Z-5.006 R7.5
N4300 X11.17 Z5.006 R7.5
N4305 X-11.17 Z-5.006 R7.5
N4310 X11.17 Z5.006 R7.5
N4315 X-11.17 Z-5.006 R7.5
N4320 X11.17 Z5.006 R7.5
N4325 X-11.17 Z-5.006 R7.5
N4330 X11.17 Z5.006 R7.5
N4335 X-11.17 Z-5.006 R7.5
N4340 X11.17 Z5.006 R7.5
N4345 X-11.17 Z-5.006 R7.5
N4350 X11.17 Z5.006 R7.5
N4355 X-11.17 Z-5.006 R7.5
N4360 X11.17 Z5.006 R7.5
N4365 X-11.17 Z-5.006 R7.5
N4370 X11.17 Z5.006 R7.5
N4375 X-11.17 Z-5.006 R7.5
N4380 X11.17 Z5.006 R7.5
N4385 X-11.17 Z-5.006 R7.5
N4390 X11.17 Z5.006 R7.5
N4395 X-11.17 Z-5.006 R7.5
N4400 X11.17 Z5.006 R7.5
N4405 X-11.17 Z-5.006 R7.5
N4410 X11.17 Z5.006 R7.5
N4415 X-11.17 Z-5.006 R7.5
N4420 X11.17 Z5.006 R7.5
N4425 X-11.17 Z-5.006 R7.5
N4430 X11.17 Z5.006 R7.5
N4435 X-11.17 Z-5.006 R7.5
N4440 X11.17 Z5.006 R7.5
N4445 X-11.17 Z-5.006 R7.5
N4450 X11.17 Z5.006 R7.5
N4455 X-11.17 Z-5.006 R7.5
N4460 X11.17 Z5.006 R7.5
N4465 X-11.17 Z-5.006 R7.5
N4470 X11.17 Z5.006 R7.5
N4475 X-11.17 Z-5.006 R7.5
N4480 X11.17 Z5.006 R7.5
N4485 X-11.17 Z-5.006 R7.5
N4490 X11.17 Z5.006 R7.5
N4495 X-11.17 Z-5.006 R7.5
N4500 X11.17 Z5.006 R7.5
N4505 X-11.17 Z-5.006 R7.5
N4510 X11.17 Z5.006 R7.5
N4515 X-11.17 Z-5.006 R7.5
N4520 X11.17 Z5.006 R7.5
N4525 X-11.17 Z-5.006 R7.5
N4530 X11.17 Z5.006 R7.5
N4535 X-11.17 Z-5.006 R7.5
N4540 X0. Z7.5 R7.5
N4545 Z-7.5 R7.5
N4550 Z7.5 R7.5
N4555 Z0. R3.75
N4560 G00 Z5.
N4565 Z15.
N4570 M09
N4575 G94
N4580 G97 S1000 M03
N4585 G00 X0. Z15.
N4590 Z5.
N4595 X4.188
N4600 Z2.183
N4605 G01 Z-2.817 F0.
N4610 G13 X18.533 Z-2.094 R9.5
N4615 X-18.533 Z2.094 R9.5
N4620 X18.533 Z-2.094 R9.5
N4625 X-18.533 Z2.094 R9.5
N4630 X18.533 Z-2.094 R9.5
N4635 X-18.533 Z2.094 R9.5
N4640 X18.533 Z-2.094 R9.5
N4645 X-18.533 Z2.094 R9.5
N4650 X18.533 Z-2.094 R9.5
N4655 X-18.533 Z2.094 R9.5
N4660 X18.533 Z-2.094 R9.5
N4665 X-18.533 Z2.094 R9.5
N4670 X18.533 Z-2.094 R9.5
N4675 X-18.533 Z2.094 R9.5
N4680 X18.533 Z-2.094 R9.5
N4685 X-18.533 Z2.094 R9.5
N4690 X18.533 Z-2.094 R9.5
N4695 X-18.533 Z2.094 R9.5
N4700 X18.533 Z-2.094 R9.5
N4705 X-18.533 Z2.094 R9.5
N4710 X18.533 Z-2.094 R9.5
N4715 X-18.533 Z2.094 R9.5
N4720 X0. Z9.5 R9.5
N4725 Z-9.5 R9.5
N4730 Z9.5 R9.5
N4735 Z0. R4.75
N4740 G00 Z5.
N4745 Z15.
N4750 M09
N4755 G14
N4760 M30
%
```

Setup Sheet - -Part 1001

Job

WCS: #0

STOCK:

  DX: 70mm
DY: 70mm
DZ: 100mm

PART:

  DX: 63.48mm
DY: 63.48mm
DZ: 94.37mm

STOCK LOWER IN WCS #0:
  X: -35mm

Total

NUMBER OF OPERATIONS: 10

NUMBER OF TOOLS: 6

TOOLS: T1 T4 T5 T6 T7 T8

MAXIMUM Z: 15mm

MINIMUM Z: -110.4mm

Operation 1/10
DESCRIPTION: Face1
STRATEGY: Unspecified
WCS: #0
TOLERANCE: 0.01mm

MAXIMUM Z: 5mm
MINIMUM Z: -2.82mm
MAXIMUM SPINDLE SPEED: 500rpm
MAXIMUM FEEDRATE: 1000mm/min
CUTTING DISTANCE: 48.39mm
RAPID DISTANCE: 57.19mm
ESTIMATED CYCLE TIME: 4s (0.9%)
COOLANT: Off

T1 D0 L0
TYPE: general turning
DIAMETER: 0mm
LENGTH: 0mm
FLUTES: 1

Operation 2/10
DESCRIPTION: Profile1
STRATEGY: Unspecified
WCS: #0
TOLERANCE: 0.01mm
STOCK TO LEAVE: 0.2mm/0.3mm
MAXIMUM STEPDOWN: 2mm
MAXIMUM STEPOVER: 3mm

MAXIMUM Z: 5mm
MINIMUM Z: -100.98mm
MAXIMUM SPINDLE SPEED: 500rpm
MAXIMUM FEEDRATE: 1000mm/min
CUTTING DISTANCE: 430.21mm
RAPID DISTANCE: 383.75mm
ESTIMATED CYCLE TIME: 30s (7.9%)
COOLANT: Flood

T8 D0 L0
TYPE: boring turning
DIAMETER: 0mm
LENGTH: 0mm
FLUTES: 1

Operation 3/10

Description: Groove1
Strategy: Unspecified
WCS: #0
Tolerance: 0.01mm
Stock to Leave: 0mm
Maximum stepover: 1mm

Maximum Z: 5mm
Minimum Z: -54.68mm
Maximum Spindle Speed: 500rpm
Maximum Feedrate: 1000mm/min
Cutting Distance: 32.18mm
Rapid Distance: 144.08mm
Estimated Cycle Time: 4s (1%)
Coolant: Off

T6 D0 L0
Type: groove turning
Diameter: 0mm
Length: 0mm
Flutes: 1

Operation 4/10

Description: Profile2
Strategy: Unspecified
WCS: #0
Tolerance: 0.01mm
Stock to Leave: 0mm
Maximum stepdown: 1mm
Maximum stepover: 1mm

Maximum Z: 5mm
Minimum Z: -110.4mm
Maximum Spindle Speed: 500rpm
Maximum Feedrate: 1000mm/min
Cutting Distance: 376.35mm
Rapid Distance: 365.31mm
Estimated Cycle Time: 27s (7%)
Coolant: Flood

T8 D0 L0
Type: boring turning
Diameter: 0mm
Length: 0mm
Flutes: 1

Operation 5/10

Description: Groove3
Strategy: Unspecified
WCS: #0
Tolerance: 0.01mm
Stock to Leave: 0mm
Maximum stepover: 1mm

Maximum Z: 5mm
Minimum Z: -107.35mm
Maximum Spindle Speed: 500rpm
Maximum Feedrate: 1000mm/min
Cutting Distance: 2167.71mm
Rapid Distance: 474.07mm
Estimated Cycle Time: 2m:16s (35.4%)
Coolant: Off

T6 D0 L0
Type: groove turning
Diameter: 0mm
Length: 0mm
Flutes: 1

Operation 6/10

Description: Part1
Strategy: Unspecified
WCS: #0
Tolerance: 0.01mm

Maximum Z: 5mm
Minimum Z: -101.18mm
Maximum Spindle Speed: 500rpm
Maximum Feedrate: 1000mm/min
Cutting Distance: 91.6mm
Rapid Distance: 212.37mm
Estimated Cycle Time: 8s (2.1%)
Coolant: Off

T6 D0 L0
Type: groove turning
Diameter: 0mm
Length: 0mm
Flutes: 1

Operation 7/10

Description: Drill2
Strategy: Drilling
WCS: #0
Tolerance: 0.01mm

Maximum Z: 15mm
Minimum Z: -102.59mm
Maximum Spindle Speed: 1000rpm
Maximum Feedrate: 250mm/min
Cutting Distance: 209.55mm
Rapid Distance: 25.63mm
Estimated Cycle Time: 51s (13.2%)
Coolant: Off

T4 D4 L4
Type: drill
Diameter: 12mm
Tip Angle: 118°
Length: 130.4mm
Flutes: 1

Operation 8/10

Description: Drill3
Strategy: Drilling
WCS: #0
Tolerance: 0.01mm

Maximum Z: 15mm
Minimum Z: -95.18mm
Maximum Spindle Speed: 1000rpm
Maximum Feedrate: 0mm/min
Cutting Distance: 2890.85mm
Rapid Distance: 185.87mm
Estimated Cycle Time: 2s (0.6%)
Coolant: Off

T5 D5 L5
Type: drill
Diameter: 8mm
Tip Angle: 118°
Length: 190mm
Flutes: 1

Operation 9/10

Description: Drill4
Strategy: Drilling
WCS: #0
Tolerance: 0.01mm

Maximum Z: 15mm
Minimum Z: -54.68mm
Maximum Spindle Speed: 1000rpm
Maximum Feedrate: 0mm/min
Cutting Distance: 2508.6mm
Rapid Distance: 90mm
Estimated Cycle Time: 1s (0.3%)
Coolant: Off

T7 D7 L7
Type: drill
Diameter: 15mm
Tip Angle: 118°
Length: 195mm
Flutes: 1

Operation 10/10

DESCRIPTION: Drill5

STRATEGY: Drilling

MAXIMUM Z: 15mm

MINIMUM Z: -14.03mm

MAXIMUM SPINDLE SPEED: 1000rpm
MAXIMUM FEEDRATE: 0mm/min
CUTTING DISTANCE: 749.11mm
RAPID DISTANCE: 51.35mm

T7 D7 L7

TYPE: drill
DIAMETER: 15mm
TIP ANGLE: 118°
LENGTH: 195mm

Generated by Inventor HSM Pro 4.0.0.032

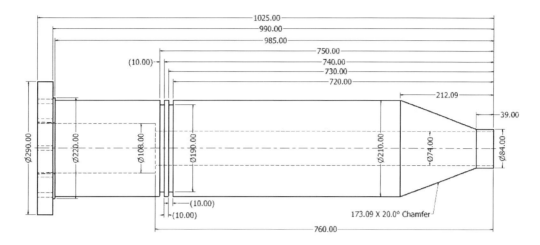

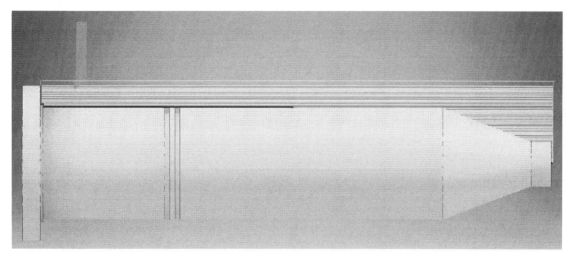

Figure 5.17 Drawing and tool path for part 1102

Program for the part 1102 is not listed here due its big size. Contact the author to provide you with download link.

Notes:

Notes:

Notes:

INDEX

3D	8, 15, 20, 110
3D data model	8
5 axis	26

A

absolute coordinate	29, 30, 83
actual position	18, 72, 74
ALU	17, 18
America National Standard Institute	34
ANSI B212.2.2002	34, 35
AP-238-STEP	8
application software	19
APT	4, 5, 10,
arithmetic-logical unit	17
ASCII	5
automatic tool change	18
Automatically Programmed Tool	4
axis motion	5, 24, 91, 93

B

Bendix	9
block	42, 46, 51, 53, 70, 71, 72, 73, 74, 75, 76, 77, 91, 93, 95, 102, 104, 105, 106
block skip	45, 103, 104, 105
Boeing	8
Bridgeport	9
bubble memory	18

C

CAD	2, 5, 8, 20
CAD/CAM	4, 8
CAD/CAM software	5, 9, 10, 61
calculations	17, 19, 29
CAM	2, 7, 8, 16, 20,
canned cycle	42, 44, 45, 79
Cartesian coordinate system	24 27, 28
cathode ray tube	18
CATIA	20
Cimatron	20
Cincinnati Milacron	6, 9
circular interpolation	17, 44, 49, 50, 51, 52, 53, 97, 100, 101, 102
clearance	34
CNC	2, 4, 5, 6, 7, 8, 9, 13, 14, 16, 17,18, 19, 20, 23, 24, 25, 26, 27, 29 ,41, 42, 44, 62, 65, 68, 72, 74, 82, 84, 86, 91, 93, 95, 96, 98, 101, 102, 104, 105, 106, 109, 110

CNC control system   14
CNC machine tools   4, 6
communications   17, 18,
Computer Aided CNC process flow   20
Computer Control Unit   17
continuous path   14, 15, 16,17
contouring   16
control logic   6
Control system   2, 3, 14, 15
    control panel   93, 104, 105
    feature   8, 29, 34, 58, 59, 68, 105
    memory capacity   6
conversational programming   8
coolant control   18
counters   18, 34
countersinks   34
CPU   6, 17, 18, 19
CRT   18
CSS   44
cutter path   8, 20
cutter radius offset   62, 64
cutting-point   34, 35
cutting-point configuration   34, 35
cycle start   86, 92, 93

D

depth of cut   42
diametrical values   28, 29, 30, 31, 32, 33
diametrical X   28
diametrical/radius values   29

diamond   34, 47, 48, 50, 52, 55, 57, 59, 61, 63, 65, 67, 70, 73, 75, 78, 81, 82, 84, 85, 87, 89, 91, 95, 97, 98, 100, 101, 103,104, 106

Direct Numerical Control   6
Distributed Numerical Control   7
DNC   6, 7
DNC   6, 7

drawing   19, 20,28, 29, 46, 48, 50, 52, 54, 55, 57, 59, 61, 63, 65, 67, 69, 70, 72, 75, 76, 78, 81, 82, 84, 85, 87, 89, 90, 92, 94, 95, 97, 98, 99, 101, 102, 104, 106, 110, 111, 115, 121, 125, 128, 132, 138, 146, 154, 161, 164, 167, 170, 176, 180, 183, 191

    surface finish   35, 84
    tolerances   34, 35
drilling operations   peck drilling   76

drilling operations reaming	14
dwell command	53

E

edge preparation	34, 35
EIA	4,7, 8, 9
Electronic Industries Alliance	4, 7
EMCO	9
ENIAC	3

F

facet size	34, 35
Fanuc	9
feed rate control	
feed per minute	90
feed per revolution	91
feedback	5,18
first digital computer	3
fixed cycles selection	8, 20, 42
flash memory	18
Flexible Manufacturing System	2
floppy disk	5, 6, 18
FMS	2
four quadrant	27, 28
fundamental concepts	23

G

G-code	4
G00 Rapid Linear Motion	46
G01 Linear Interpolation	47
G02 Circular Interpolation Clockwise (CW)	49
G03 Circular Interpolation Counter Clockwise (CCW)	51
G04 Dwell	53, 54
G20 Inch Units	55
G21 Millimeter Units	56
G28 Return to Home Position	58
G29 Return from Home Position	58
G32 Single Thread Cycle	60
G40 Tool Nose Radius Compensation Cancel	66
G41 Tool Nose Radius Compensation Left	62
G42 Tool Nose Radius Compensation Right	64
G54-G59 Select Work Coordinate System	68

G70 Finishing Contour Cycle	70
G71 Rough Turning Contour Cycle	71
G72 Rough Facing Contour Cycle	73
G74 Peck Drilling Cycle	76
G75 Grove Cycle	77, 78
G76 Threading Cycle	79
G90 Absolute Programming	81
G91 Incremental Programming	83
G96 Constant surface speed	84, 85
G97 Constant Spindle Speed	86
G98 Feed Rate Per Time	88
G99 Feed Rate Per Revolution	90
General Electric	6, 9
geometry	3, 4, 62
Giddings	9

H

Haas	9
hard drive	18
hard wired	6
hexagon	34
home position	58, 59, 60, 61, 66, 78, 105

I

I/O	17, 18
incremental coordinate	31, 82
input/output	17, 18
inscribe circle	34, 35
interface	2, 17, 18, 19
interlacing to devices punched tape	6
Intermediate point	58, 59
International Standard Organization	7
interpolation	3, 16, 17, 44, 47, 48, 49, 50, 51, 52, 52, 53, 54, 55, 56, 57, 58, 60, 61, 62, 64, 66, 67, 68, 69, 77, 78, 79, 82, 83, 84, 85, 86, 88, 89, 90, 91, 92, 93, 94, 95, 96, 97, 98, 99, 100, 101, 102, 103, 105, 106,
Inventor HSM	19, 110, 114, 120, 124, 127, 131, 137, 145, 153, 160, 163, 166, 169, 175, 179, 182, 190
ISO 1832-1991	34
ISO 6383	7
ISO STEP	8

L

lathes	27, 88, 90, 96, 98

LCD		18
left hand		35
Lewis		9
limit (end) switch		18
linear interpolation		16, 44, 47, 48, 50, 52, 53, 54, 56, 57, 60, 61, 64, 66, 67, 77, 78, 82, 83, 84, 85, 86, 88, 89, 90, 91, 92, 93, 94, 96, 97, 98, 99, 100, 101, 102, 103, 105, 106
linear path		16
liquid crystal display		18

M

M code		8, 42, 44, 105,
M00 Program Stop		91, 92, 93
M01 Optional Program Stop		93, 94
M02 Program End		95
M03 Spindle Clockwise Rotation (CW)		96
M04 Spindle Counter Clockwise Rotation (CCW)		98
M05 Spindle Stop		99
M08 Coolant Start		100
M09 Coolant Off		101
M30 Program End and Reset to the Beginning		102
Machine Control Unit		17
machine zero		58
machine zero	intermediate point	58,59
machining holes	peck drilling	76, 77
manual		2, 3, 8, 19, 42, 88, 90, 101
manual programming		8
Mastercam		20
MCU		17
memory		5, 6, 17, 18
microprocessor		6
Milltronics		9
Mitsubishi		9

N

NC		2, 3, 4, 5, 6, 7, 8, 18, 19
network interface		18
neutral		35
numerical control		1, 2, 3, 5, 6, 7,
numerical control	advantages	2, 80
numerical	definition	4

control

O

octagon		34
offsets		8
Okuma		9
operating system		18, 19
optional stop		93
origin		28, 29, 30, 31, 32, 82, 83

P

parallelogram		34
parameters		35, 44, 49, 51, 72, 74, 79, 80
part drawing		19, 28
part program		9, 18, 19, 20, 21
pattern of holes		2
peck drilling		76, 77
pentagon		34
planes		2
point to point		14, 15
position compensation	Z-axis	30
position register commands	milling	3, 10, 14, 15, 24, 26, 27, 27, 29, 34, 41, 42, 44, 47, 48, 50, 52, 55, 57, 59, 61, 62, 63, 65, 67, 70, 71, 73, 75,78, 80, 81,
position register commands	turning	82, 84, 85, 87, 89, 91, 95, 97, 98, 100, 101, 103, 104, 106, 109, 110,114, 120, 131, 153, 166, 175, 179, 182,
process flow		19, 20
profile		28, 29, 31, 61, 62, 64, 70, 71, 72, 73, 74, 75
program		2, 4, 5, 6, 7, 8, 9, 10, 16, 18, 19, 20, 21, 24, 29, 42, 44, 45, 46, 47, 48, 49, 50,51, 52, 53, 54, 55, 56, 57, 58, 59, 60, 61, 62, 63, 64, 65, 66, 67, 68, 69, 70, 71, 72, 73, 74, 75, 76, 77, 78, 79, 80, 81, 82, 83, 84, 85, 86, 87, 88, 89, 90, 91, 92, 93, 94, 95, 96, 97, 98, 99, 100, 101, 102, 103, 104, 105, 106, 110, 191
program end		45, 47, 49, 51, 53, 54, 56, 57, 60, 62, 64, 66, 68, 69, 71, 73, 76, 77, 79, 81, 83, 84, 86, 88, 90. 91, 93, 94, 95, 96, 97, 99, 100, 102, 103, 105, 106
program coordinates		19
program identification	program number	47, 48, 50, 52, 54, 56, 57, 60, 61, 63, 65, 67,69, 71, 73, 75, 77, 78, 81, 82, 84, 85, 87, 89, 91, 92, 94, 95, 97, 98, 100, 101, 103, 104, 106
program stop		91, 92, 93, 94, 101

program verification	20
program verification thread cutting	81, 86, 87
program zero lathes	27, 88, 90, 96, 98
program zero selection methods	17, 18
programming language	4
prototype	21
PTP	14, 15
punched tape	6
punching card	2

Q

quadrants	27, 28

R

radius programming	29
radius values	29, 79
rakes	34
RAM	18
Random Access Memory	18
rapid positioning tool path motion	4
Read Only Memory	18
rectangle	34
rectangular Cartesian coordinate system	24
right hand	24, 34, 35, 47, 48, 50, 52, 55, 57, 59, 61, 63, 65, 67, 70, 73, 75, 78, 81, 82, 84, 85, 87, 89, 91, 95, 97, 98, 100, 101, 103, 104, 106
right-hand coordinate system	24
right-hand rule	24, 26
R-level selection	18
ROM	18
rotation axes	16
round	34
RS232	18
RS274/NGC	4
RS274D	4, 8

S

sequencer	17
serial communications	18
servo	3, 5, 17, 18
servo drive control	17, 18

setup	19, 20, 24, 55, 56, 63, 65, 68, 82
setup sheet	19, 20, 110, 113, 118, 123, 127, 131, 136, 144, 151, 160, 163, 166, 169, 175, 179, 182, 188
shape	2, 8, 16, 34, 35, 47, 48, 50, 52, 55, 57, 59, 61, 63, 65, 67, 69, 70, 73, 75, 78, 81, 82, 84, 85, 87, 89, 91, 95, 97, 98, 100, 101, 103, 104, 106
Siemens	9, 20
Siemens NX	20
simultaneous control	16
size	3, 34, 35, 62, 69, 79, 80, 84, 85, 92, 93
SolidCAM	20
spindle control constant surface speed	44, 84, 85, 85, 86, 87, 99, 191
spindle speed control	17, 18
square	34, 78, 104
storage	5, 6, 17, 18,
storage interface	18
SurfCam	20
system bus	17, 18

T

tapping	14
target position	79
thickness	34, 35
threading	14, 44, 61, 79
depth of thread	61, 62, 79, 80
G76 cycle	61, 79,
retract from thread	24
tapered thread	79
timer	18
tolerance class	34
tool nose radius offset	62, 64
tooling	19, 20
tools	3, 4, 5, 9, 20, 24, 34, 63, 65, 71, 73, 75, 81, 87
triangle	34, 35
trigon	34
turning	27, 29, 34, 42, 44, 47, 48, 50, 52, 55, 57, 59, 61, 62, 63, 65, 67, 70, 71, 73, 75, 78, 80, 81, 82, 84, 85, 87, 89, 91, 95, 97, 98, 100, 101, 103, 104, 106, 109, 110, 114, 120, 131, 153, 166, 175, 179, 182
turning CNC machine	27
turning tools	62
type	2, 7, 9, 16, 18, 21, 34, 35, 47, 48, 50, 52, 55, 57, 59, 61, 63, 65, 67, 70, 73, 75, 78, 79, 80, 81, 82, 84, 85, 87, 88, 89, 90, 91, 95, 97, 98, 100, 101,

202

103, 104, 106, 113, 114, 119,120, 123, 124, 127, 131, 137, 144, 145, 152, 153, 160, 163, 166, 169, 175, 179, 182, 188, 189, 190

V

verify   19, 20, 21, 31
virtual machine   19, 20

W

wear offset   adjustment   92, 93
Whirlwind   3
word address format   word   4, 8, 42
work coordinate system   68
work offsets   G54-G59 commands   68
workpiece   24, 27, 28, 29, 31, 46, 58, 59, 68, 72, 74, 100

Y

Yasnak   9

Z

ZX coordinate plane   27

Notes:

Made in the USA
Monee, IL
19 December 2023

49934553R00125